Statistical Monitoring of Clinical Trials

Lemuel A. Moyé

Statistical Monitoring of Clinical Trials

Fundamentals for Investigators

With 69 Figures

Lemuel A. Moyé
University of Texas
Health Science Center at Houston
School of Public Health
Houston, TX 77030
USA
moyelaptop@msn.com

Library of Congress Control Number: 2005931408

ISBN-10: 0-387-27781-1 e-ISBN 0-387-27782-X
ISBN-13: 978-0387-277813

Printed on acid-free paper.

Printed in the United States of America. (MVY)

9 8 7 6 5 4 3 2 1

springeronline.com

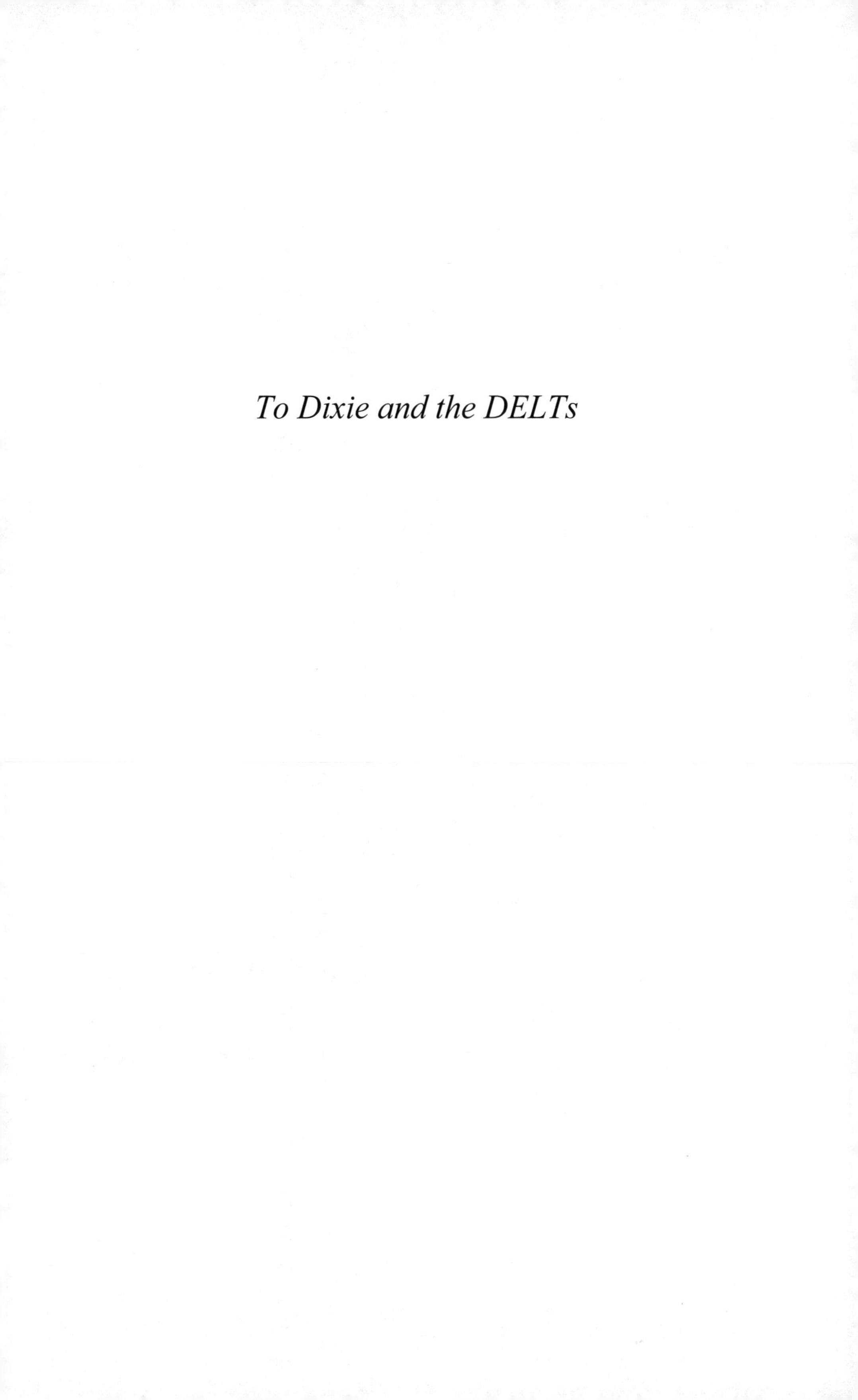

To Dixie and the DELTs

Preface

A preface is an opportunity for you and me to share an amiable conversation before the serious work starts. If you give me a moment, I will share with you my motivations for writing an introductory text about the statistical monitoring of clinical trials, a staple of modern research efforts in heatlhcare.

I am pleased to have been involved in clinical research for eighteen years. Many of my efforts focused on preparations for and presentations to Data Monitoring Committees (DMCs), each of which was tasked with overseeing the conduct of a particular clinical study. During these activities, I have spoken with many clinicians about the epidemiology and biostatistical foundation of this mode of clinical research.

In my experience, nothing confuses a DMC member as do these so-called "stopping rules" for monitoring the conduct of a healthcare research study. The idea of prematurely ending a study makes intuitive sense to the clinical members of the committee. The rules themselves with their arcane terminology are the problem. Descriptions of "group sequential procedures" and "stochastic curtailment" provide no useful handholds for the clinician working to understand this slippery but essential subject. The fact that neither medical school nor residency curricula discuss any of the details of these procedures is one possible explanation for the continued lack of understanding among clinicians. In general, the non-statistical members of newly conceived DMCs in the 21st century are just as confused about statistical monitoring guidelines as were their clinical predecessors who sat on DMCs in the 1980s.

A major reason for this continued confusion is that clinical investigators, although blessed with the motivation to do research, commonly do not have strong mathematical backgrounds. Although many have worked hard to develop the basic understanding of epidemiology and biostatistics necessary to be an effective investigator, the underlying mathematical details of commonly used monitoring procedures as frequently presented remain beyond the scope of their training.

Of course, the statistical literature has much to say on the subject of monitoring rules in clinical research. Beginning with the manuscripts of Armitage and Wald in the 1940s, the statistical treatment of this topic slowly expanded until the late 1970s, when it exploded. The recognition of the importance of the monitoring of clinical research, in concert with the complexity of the underlying mathematics has attracted the best and the brightest of biostatisticians. Their devotion to the study of the underlying mathematical structure of monitoring procedures has resulted in a body of knowledge that is both evolutionary and illuminating. However, because it tends to be scripted in the technical and exclusionary language of advanced mathematics, the writing tends to enlighten only the sophisticated analyst.

The text by Christopher Jennison and Bruce Turnbull [1] is a fine example of a comprehensive treatment of a difficult statistical subject.

The technical writing style that has been implemented in the field of "interim monitoring" should come as no surprise. However, work in this area, propelled forward by the strong rowing of capable statistical theorists, can leave the clinical investigator behind in its wake. Required to apply complex processes that they do not understand, the clinical investigator commonly finds little introductory material available. In addition, clinical researchers with no quantitative background have difficulty communicating with biostatisticians or experienced trial methodologists who have much experience but little time to explain these issues to their inexperienced colleagues. Thus, investigators who wish to learn about these mathematical procedures are hard pressed to identify readily understandable source material.

The purpose of this text is to fill that gap. If you know nothing about monitoring guidelines in clinical trials, then this book is for you.

I have chosen to begin this book with a brief history of monitoring rules in clinical research. Although this is the first chapter in this book, it needn't be the first chapter that you read. Being nontechnical, it might be most useful to view its contents as a pleasant oasis in a desert of more complicated discussion. Its considerations of the interactions between scientists serves to convey something about the people who were involved in these important historical efforts. The observation that the epidemiologist Bradford Hill suffered from tuberculosis years before he helped design an early clinical trial to study this disease may be a mere curiosity to some; to others it helps to explain his intellectual fortitude in working with skeptical clinicians.

For the same reason, I have broken up some of the technical arguments that appear in later chapters with an occasional vignette. As my students frequently remind me, it is best to have a joke close by when discussing anything mathematical.

I must confess that this is not a book about the operation of DMCs. That material has been very nicely developed in *Data Monitoring Committees in Clinical Trials: A Practical Perspective* by Susan Ellenberg, Thomas Fleming, and David DeMets (John Wiley & Sons, Ltd., West Sussex, 2002). Their text is very broad in scope, focusing on the DMCs evolution and contemporary operation. Our focus here is on statistical monitoring procedures that these DMCs devise and utilize, not on the DMCs themselves.

One final note. An important segment of the current clinical investigator population is comprised of women. Therefore, I have alternated the use of gender in the hypothetical illustrations offered by this text. Although this is the most illustrative and the least exclusionary approach, it does require mental alacrity on your part as the genders change from example to example.

Lemuel A. Moyé
The University of Texas
School of Public Health
Houston, Texas
July, 2005

References

1. Jennison C, Turnbull BW. (2000). *Group Sequential Methods with Applications to Clinical Trials*. New York. Chapman & Hall/CRC.

Acknowledgments

My understanding of monitoring procedures was not generated by reading complicated treatises, but instead was forged in vibrant and sometimes fiery discussions during Data Monitoring Committee meetings. The distinguished members serving on these panels deserve much of the credit for the educational material this book contains. Special thanks goes to the members of the University of Texas School of Public Health Coordinating Center for Clinical Trials, and to its senior members both past and present. Mort Hawkins ScD, Barry Davis, MD, PhD, and Robert Hardy, PhD. served as the guide rails for me, keeping me on the right track when I tended to veer too far in one direction or the other.

I also acknowledge the members of the Houston SPOTRIAS team. In this effort, we investigators have faced numerous challenges in working to devise useful monitoring guidelines for small pilot research projects, and in doing so, have created through debate and discussion many good ideas. Special thanks to Dr. James Grotta, Andre Alexandrov and Louis Morgenstern for their roles in these conversations.

John Kimmel, the reviewing editor, and several chapter reviewers have provided good questions, challenges, suggestions, and additional references. Their help has improved this book's structure.

Nidhi Rohatgi, my graduate teaching assistant, provided invaluable assistance in clarifying the text and in copyediting. Nidhi was especially helpful in pointing out the need for me to untangle what had become relatively impenetrable tracts of some chapters. Dr. Claudia Pedroza provided fine support for the discussion of Bayesian monitoring guidelines.

Finally, my dearest thanks go to Dixie, my wife, on whose personality, character, love, and common sense I have come to rely, and to Flora and Bella Ardon, whose continued emotional and spiritual growth reveals anew to me each day that, through God, all things are possible.

Lemuel A. Moyé, MD, PhD
University of Texas
School of Public Health
Houston, Texas, USA
July, 2005

Contents

Introduction

Statistical monitoring procedures are the body of computations that aid clinical investigators in determining if a research program should be suspended prematurely. Specifically, these guidelines are used to guide the complex decision to end a clinical study if the investigation is very likely to produce either (1) an early positive benefit, (2) an early indication of harm, or (3) a neutral effect at the time the study is scheduled to end (expressed as stopping for "futility"). Research scientists and members of clinical trial oversight committees rely upon these procedures, colloquially expressed as "stopping rules", but more correcting described as "monitoring guidelines".

Although clinical investigators accept the application of statistical and epidemiologic principles in clinical research, the procedures used to terminate clinical studies often appear opaque to the statistically naïve investigator. Nevertheless, these guidelines have become ubiquitous in healthcare research. In 1998, the Office of the Inspector General of the Department of Health and Human Services mandated that the Food and Drug Administration (FDA) and the National Institutes of Health (NIH) develop such procedures and standards for U.S. trials. In response, the NIH has generated policies to require safety monitoring plans for all phase III NIH-funded studies, and the FDA has issued a draft guidance document on the establishment and operation of the committees that perform such monitoring. In addition, the Institutional Review Boards (IRBs) that govern the ethical conduct of clinical investigation at many research centers developed their own sets of instructions for the application of oversight procedures. These monitoring responsibilities reside in the Data Monitoring Committees (DMCs) of the individual clinical research projects.

This new requisite for formal statistical monitoring of clinical research places clinical investigators in a dilemma. As researchers in a study, they have to satisfy the monitoring requirements of their institutional review board. Alternatively, if they are members of a DMC, then their input into the discussions that calibrate the statistical monitoring device of the study is required. However, these investigators are commonly ill equipped to deal with the issues of modern statistical monitoring of clinical trials. Thus, they are unable to fruitfully engage in the discussion, development, or defense of the use of these tools.

Well-motivated, but statistically unsophisticated clinical investigators can learn the correct use and interpretation of these monitoring procedures when provided with a learning tool that informs them in clear language. This tool would allow them to steadily increase their knowledge of, experience with, and intuition about these procedures. *Statistical Monitoring of Clinical Trials: Fundamentals for Investigators* is this tool. Specifically, it provides the discussion of these statistical devices that clinical investigators need, representing a user-friendly introduction to monitoring procedures for these scientists. These essential statistical considerations are rarely taught in introductory biostatistics or medical statistics classes.

Chapter One of *Statistical Monitoring of Clinical Trials: Fundamentals for Investigators* provides an overview of the evolution of monitoring procedures in clinical research. Randomized, blinded controlled clinical trials, available for only sixty years, are a relatively new tool in clinical investigation, and remain controversial. The ethical concerns raised by this investigational methodology have called for the interim monitoring of these studies. This demand in turn has generated a relatively new application for Brownian motion, one completely unforeseen by its progenitors, including Albert Einstein.

Chapter Two provides a review of the basic statistical thought process required in clinical research and directly applicable to interim monitoring. The set of circumstances that permit one to generalize the results from a single small sample to a population of thousands or millions of subjects has direct bearing on the successful application of statistical monitoring of clinical trials. These situations and their limitations are discussed in detail. In addition, the foundation principles of statistical hypothesis testing, confidence intervals, and the Bayes approach are each described.

Chapter Three develops the elementary principles of probability that are required to understand the principles behind the interim review of clinical research results. The differences between subjective and objective probability are discussed, and the roles of each in the statistical monitoring of clinical trials are explained. In addition, the concept of probability as an area under a curve is illuminated, with special emphasis given to the normal distribution. Finally, elementary examples of the use of probability for the early termination of a clinical research effort are provided. Chapters Two and Three provide the foundation for the rest of the text.

Chapter Four addresses the need for monitoring procedures in clinical research. This chapter lays out for the clinical scientist the problems that arise when one attempts to use traditional hypothesis testing procedures to draw conclusions about a clinical study's interim results. It provides, through the use of discussion and examples, the elaboration the clinical scientist needs in order to develop insight into the basic behavior of statistical monitoring tools. Investigators have become familiar with the idea of a test statistic's location (i.e., whether the test statistic is greater than 1.96). In this chapter, that notion is supplemented with the observation that a test statistic follows a particular path to arrive at its current location. An examination of that path's properties reveals new information that can provide accurate predictions of the test statistic's location in the future. This concept is new to most clinical investigators, and is elaborated in detail without heavy reliance on mathematics. It is here that the link between Brownian motion and clinical monitoring procedures is motivated.

Capitalizing on the insight provided in Chapter Four, Chapter Five introduces the basic group sequential approach of Pocock and O'Brien–Fleming, followed by discussions of the Haybittle–Peto and Lan–DeMets derivatives. The triangular designs popularized by Whitehead are briefly discussed. Chapter Six develops conditional power in a way that illuminates the circumstances in which a clinical trial may be stopped early for a beneficial finding based on a "look forward" approach.

Chapter Seven describes the use of monitoring procedures to identify harmful effects of the tested intervention. This is a natural introduction to the current use of asymmetric monitoring procedures. In addition, the problem of deciding to discontinue a study because of an unanticipated finding in one of several safety measures is developed. The many unexpected safety considerations that can arise during the study's execution amplify the importance of this issue. This chapter also introduces the notion of stopping a clinical trial early due to "futility".

Chapter Eight provides an introduction to the use of monitoring procedures using the Bayes paradigm. Each chapter ends with a relevant problem set.

This book can serve as a reference text for clinical scientists at all levels of training, being especially useful for healthcare graduate students and junior physician-scientists. Its readers require basic college algebra, plus one course in healthcare statistics. Its contents are of interest to students attending medical schools, graduate schools with an emphasis in healthcare research, and schools of public health. In addition, the contents of *Statistical Monitoring of Clinical Trials: Fundamentals for Investigators* are applicable to workers in health departments, private institutes, and government regulatory agencies. This book is also useful for judges who, not uncommonly, have to learn about the ethical conduct (and, therefore, the ethical monitoring) of clinical research efforts.

This text's incorporation of background material as well as in-depth discussion requires some guidance for its optimal use. There are several sections in Chapters 5, 6, and 7 which have a "*" in their title, signifying that the material is more challenging for students with a weak background in probability. In addition, the appendices, providing some in-depth mathematical development, can also appear formidable to a student with one background course in statistics.

Therefore, this book may be successfully used as the basis for a basic, introductory course on monitoring rules in clinical trials by focusing on Chapters 1 through Chapter 7, ignoring (1) all of the starred sections in Chapters 5, 6, and 7 and (2) the contents of Appendices A through D. However, those with a stronger mathematics background, after reviewing the historical introduction, can move directly to Chapter 4 and proceed through Chapter 8, covering the details of Appendices A through E as needed.

One caveat. Healthcare researchers regardless of their level of mathematical sophistication, should spend some time in Chapter Two, which discusses the statistical reasoning process in medicine. The experience of the author is that, without this review, many researchers unfortunately use statistical monitoring procedures as a tool to identify the "smallest p-value the quickest way" leading to important setbacks in the development of both research programs and research careers.

1

Here, There be dragons....

What will clinical research look like in the year 2065?

The veil of uncertainty shielding our view of the future blocks any detailed response to this provocative question. We might attempt the answer that "in 2065, research will strike the right balance between compassion on the one hand, and the needs of investigational science on the other, their interaction being governed by a overarching ethic." However, this is more of a hope than an observation. Try as we might, we cannot reliably comment on the methodology to be implemented in the mid-21st century.

Just as, we have only the dimmest view of clinical investigation 60 years from now, early clinical trialists working in the 1940s could not imagine what clinical investigation would look like at the end of the 20th century. In the years following World War II, clinical trials fought for acceptance and respectability, struggling to take root in a soil often poisoned by cultural resistance. Many researchers in the 1940s hoped that the "clinical trial" would die a quick death, rubbed out by the ethical dilemmas raised by its use of randomization and treatment blinding.

At that time, linking the random movement of a pollen grain to observations of a clinical trial's treatment effect would have been dismissed as fanciful science fiction. The ideas of Brownian motion were too abstract to be helpful; they were too far removed from any recognizable structure on the clinical research map. These mysterious mathematical tools, like the unknown reaches of the earth located far from Europe on an ancient map, would have simply been stamped with the admonition, "Here, there be dragons."

The following preliminary discussion will etch out the brief history of clinical trials and Brownian motion as these separate fields drifted toward each other. We will see that the mixture of these diverse disciplines has been predictably unpredictable, an observation that we must keep in mind as we plan the trials of the 21st century.

1.1 Clinical Investigation Before the 1940s

Clinical investigation has been a human endeavor for over two thousand years. The most common building block in the edifice of health study is the case report. A case report is a summary of a single patient's findings and the communication of those findings to the medical community. A case series is a collection of case reports, linked together by a common thread (e.g., all of the patients were seen by the same doctor, or each of the patients was exposed to the same agent, e.g., quinine).

It is easy to understand how the growth of general medical knowledge has been propelled by the use of case reports. The delivery of healthcare has been governed by the interaction between a single, concerned, responsible provider and his patient. This relationship is private and privileged. However, it has historically been conducted in isolation, by physicians and nurses widely separated from each other. The idea of a community was well established. However, the concept of a medical community (i.e., a collection of practitioners who worked together to jointly expand their knowledge base) was one that took many generations to develop.

Therefore, medical care was delivered for hundreds of years by practitioners, who, working alone with incomplete knowledge, made decisions that directly affected the lives of their patients, and indirectly, their patients' families and communities. The one, natural learning tool these physicians could use was the active sharing of their experiences among themselves. This served to expand their expertise, suggest alternative approaches to healthcare, and extend their knowledge. This shared experience is at the heart of the case report.

The core thesis of this approach was best captured by Celsus (circa A.D. 25) [1], who stated that "Careful men noted what generally answered the better, and then began the same for their patients." For the next 1900 years, advances in clinical medicine occurred through the combined use of careful observations, clear recorded descriptions, and deductive reasoning. The discovery that gunshot wounds could be healed without the application of burning hot oil [2] demonstrated that a case report-style observation could uncover new information and overturn prior, erroneous principles in medicine. When medical journals began to appear, the primary medical information that they dispersed was that of the case report.* Those physicians who had more exposure and experience with a medical issue compiled their case reports together into a case series that they would publish. This continues to this day. Examples are diet drugs and heart valve disease [3] and radiation poisoning [4].

However, case reports have well-established difficulties. Although they reflect very clear and honest observations, the degree to which a single case report represents a general phenomenon in the population can be subject to debate. Even though they are useful, the variability of observations across patients makes it difficult to assess whether one patient's findings summarized in a case report can be easily translated to others.

However, what the case report and essentially all investigative mechanisms in medicine hope to illuminate, by examining both the environment (e.g.,

---

* One of my favorites is an 1822 issue of *Lancet*, whose feature article was titled, "The biggest hernia that I have ever seen in a shipyard worker".

exposure to a toxin or a potential cure) and the patient's response, is the true nature of the exposure–outcome relationship. This true nature could be simply an association, or it could be causal.

An association is the coincidental occurrence of an exposure and an outcome. Its recognition (e.g., the relationship between coffee drinking and pancreatic cancer) typically does not require direct action by the medical community. A causal relationship, on the other hand, signifies that the exposure excites the production of the outcome. This more powerful, directed relationship incites the medical and regulatory communities to action. For example, the conclusion that exposure to citrus fruits reversed the symptoms of scurvy incited action by the British navy to mandate the storage of fresh fruit in the provisions of its crews for long sea voyages [5]. On the other hand, links between the use of cutting and bleedings and the remission of yellow fever were merely associative. Thus, when we as physicians examine a case report's details, we sift through the provided clinical descriptions in order to discern if the relationship between the exposure and the outcome is either causative or associative.

Epidemiologists are specialists who identify the determinants or causes of disease. They have developed criteria that would be useful in ascertaining whether an exposure causes (i.e., excites the production of) the disease. Elaborated by Sir Austin Bradford Hill [6], these tenets are based on a common sense approach to determining causality and are remarkably free from complicated mathematical arguments. These criteria acknowledge that more disease cases in the presence of the risk factor than in its absence raise a causal suspicion. In addition, determining that greater exposure (either by dose or duration) to the risk factor produces a greater extent of disease amplifies our sense that the exposure is controlling the disease's occurrence and/or severity. These two features are important characteristics of a cause–effect relationship.

Other questions posed by Hill permit us to explore the "believability" of the relationship. Is there a discernible mechanism by which the risk factor produces the disease? Have other researchers also shown this relationship? Are there other examples that help us to understand the current exposure–disease relationship? The nine precise Bradford Hill criteria are: (1) strength of association, (2) temporality, (3) dose-response relationship, (4) biologic plausibility, (5) consistency, (6) coherency, (7) specificity, (8) experimentation, (9) analogy. These are well elaborated in the literature [7].

Diligent attempts to determine whether specific case reports and case series can satisfy these causality criteria continue to provide invaluable service to patients and communities. The link between methylmercury exposure and birth defects in communities surrounding Minamata Bay, Japan, [8], and the establishment that thalidomide was the cause of the birth defects phecomelia and achondroplasia [9] are just two 20th century examples of the ability of case reports and case series to establish causal relationships that produced public health action. The identification of (1) the relationship between tick bites and Lyme disease, and (2) the link between new illnesses among postal workers and anthrax exposure in 2001 are recent examples of their continued value.

1.2 Limitations of Case Reports

Although medical knowledge has progressed through the sensitive and intelligent use of case reports and case series, there is no doubt that the illumination provided by these investigational tools is also profoundly limited. There are four major criticisms of the value of case reports and case series in determining the causal nature of an exposure–disease relationship. They are that (1) case reports and case series do not provide quantitative measures of the relationship between an exposure and a disease, (2) case reports do not always rule out other competing causes of disease, (3) case reports are subject to biases of selection (i.e., the manner in which the case report was selected may make it unreasonable to believe that its occurrence reflects an important finding in the population), and (4) measurements made in the case report may be nonstandard. These limitations reduce the contribution of case reports to our understanding of the exposure–disease relationship.

One of the most remarkable deductive failures of case reports was their false identification of the effects of cardiac arrhythmia suppression [10]. In the 1970s, considerable attention was provided to the potential of new therapies (specifically, the drugs encainide, flecainide, and moritzacine) for the treatment of dangerous ventricular arrhythmias. It was believed that these new drugs would be more effective and produce fewer side effects than the traditional, poorly tolerated medications. The effectiveness and safety of these newer drugs were examined in a collection of case series. At first, only the sickest patients were given the new therapy. When these patients survived, the investigational drug was credited with saving the patient's life. However, if the patient died, then the patient was commonly deemed "too sick to be saved" and the drug was not debited for the death.

Based on these observations, despite some opposition, a consensus developed in the cardiology community that patients with arrhythmias would benefit from the use of these new drugs. After a period of intense deliberation, the Federal Food and Drug Administration (FDA) approved the new antiarrhythmic agents. As a consequence of this approval, physicians began to prescribe the drugs not just to patients with severe rhythm disturbances, but also to patients with milder arrhythmias. This new use was consistent with the growing consensus that these drugs would be beneficial in blocking the progression of dysrhythmia from mild heart arrhythmias to more serious rhythm disturbances.

Only after the drugs were approved and on the market was a study carried out that incorporated a control group and the use of randomization. This trial, called CAST (Cardiac Arrhythmia Suppression Trial), demonstrated that, not only did the new therapies not save lives, but their use caused excess mortality [11]. The findings from CAST, demonstrated the lethality of medications whose safety had been "demonstrated" by case series.

1.3 Genesis of the Clinical Trial

By the 1940s, the limitations of the case series as an investigational tool in medicine were evident. However, the evolution of this tool into a device resembling a clinical trial required the patient efforts of the epidemiologist Sir Austin Bradford Hill.

A clinical trial is a medical experiment that is carried out in a unique research setting that must be carefully constructed. The previous section discussed the complicated series of arguments that an investigator must go through in building a causal argument. The clinical trial is the research environment in which many of these properties of the causal argument are already embedded. Upon the beginning of the clinical trial's execution, the only missing feature of the causal argument is the strength of association. This final component is provided by the execution of the study.

Specifically, in a well-designed and well-executed clinical trial, the simple demonstration of a clinically and statistically significant strength of association between the randomly allocated intervention and the prospectively defined primary analyzes is all that is necessary to demonstrate the causal nature of the relationship. This very special situation can only be successfully constructed with (1) a clear statement of the clinical question, (2) a simultaneous focus on epidemiological and biostatistical principles, and (3) disciplined research execution. There are several comprehensive references that discuss in detail the methodology of clinical trials [12,13,14].

The 1930s was a cauldron of new ideas for clinical research. The United Kingdom Medical Research Council's (MRC) Statistical Council and Statistical Research Unit was organized in 1927 [5]. One of its responsibilities was to design and conduct clinical trials in order to investigate promising treatments for modern diseases. The council was adaptive and flexible, opening itself to new and exciting research ideas. One innovative concept was the incorporation of several investigators dispersed throughout a country, all following the same protocol into one research effort. This was the early model for what we now call a multicenter study.

In the mid-1940s, the MRC had the opportunity to evaluate the effect of a new therapy, streptomycin, as a possible treatment for tuberculosis. Streptomycin was a new antibiotic that had not yet demonstrated its effectiveness in clinical experiments. Although it was relatively plentiful in the United States, its availability was limited in impoverished post-war England. The resulting study, conducted by the MRC, was to become the template for the modern clinical trial.

Bradford Hill was asked to design this study. Being both an epidemiologist, as well as a patient who had tuberculosis as a youth,* he held a special appreciation of the complexity of the work required to conclude that streptomycin would be safe and effective for this disease. Hill wished to develop a research paradigm that would produce a clear and unbiased assessment of the effects of the antibiotic. Beginning with the established notion of an experiment in which the researcher has control over the use of an intervention,† Hill successfully argued for three features of the study that were not commonly used in clinical experiments at that time.

---

* Hill himself had contracted tuberculosis as a young man. He survived a lung abscess, artificial pneumothorax, and a two-year hospitalization twenty-five years before his pivotal streptomycin study.

† An experiment in which the researcher has control of the intervention is different from an observational study, where the investigator has no control of the intervention. An example of an observational study would be that of John Snow's evaluation of the effect of the source of water on the occurrence of cholera, in which the subjects chose their water source.

These were (1) a control group, (2) an external rather than an internal method of selecting the therapy for each individual patient, and (3) blinding, or a procedure to mask both patients and physicians to the identity of the therapy to which any particular patient was assigned [15]. The modern clinical trial emerged from the first attempts to apply these innovations [16].

It is these three features that, in combination, differentiate the clinical trial from other forms of clinical investigation. However, the incorporation of the use of a control group, the random allocation of therapy, and blinding, so essential to the transformation of the clinical experiment into a modern clinical trial, was fraught with controversy. Hill's proposal for their incorporation produced dissension among the clinicians involved in this tuberculosis study. Before we discuss the strong reactions of the research and medical communities to these devices, a reaction that grew to require the need to monitor these studies, we must say a few words about these tools and their intended purposes.

1.4 The Requirement for Control

In the 1940s, the need for a control group was not self-evident to clinical investigators, and it was still common to research potentially new therapies without having patients as comparators. An example was the evaluation of penicillin, in which many of the early studies were conducted without a control group.

There were two main justifications for the absence of control groups in clinical research. The first was the belief that, when the treatment effect was large, then a comparison group would be unnecessary. The second was an ethical one; withholding an experimental treatment was unjustified and harmful when the natural history of the disease (e.g., tuberculosis) was associated with profound morbidity and mortality.*

In this environment, Hill's argument for the inclusion of a control group was not well received by the clinicians who would carry out the study. Those who believed that streptomycin could only have a beneficial effect argued forcefully against the need for a control group. These investigators knew the natural history of tuberculosis; including a comparator group would not substantially add to the body of knowledge concerning the fate of these ill patients. On the other hand, streptomycin's effects were not complete unknowns because the drug had already been partially evaluated in the United States. Why, they asked, withhold a therapy from

---

* This idea of control group was turned on its head in the Tuskegee syphilis experiment, in which a known, effective therapy was deliberately withheld. For forty years between 1932 and 1972, the U.S. Public Health Service (PHS) conducted an experiment on 399 African-American men in the late stages of syphilis. These men, for the most part illiterate sharecroppers from one of the poorest counties in Alabama, were never told what disease they were suffering from or of its seriousness. Informed that they were being treated for "bad blood," their doctors had no intention of curing them of syphilis. The data for the experiment was to be collected from autopsies of the men, and they were thus deliberately left to degenerate under the ravages of tertiary syphilis—which can include tumors, heart disease, paralysis, blindness, insanity, and death. "As I see it," one of the doctors involved in the study explained, "we have no further interest in these patients until they die." Additional information is available from http://www.infoplease.com/ipa/A0762136.html.

ill patients (likely to die using the standard treatment of care), that was probably safe and could help them?

Hill countered that streptomycin had been incompletely studied to date and must be considered to have unknown effects. If, he argued, the safety and efficacy of streptomycin had already been established, there would be no need to re-evaluate the drug in England.

This scenario was especially disturbing to clinicians, because one of the worst things that they could do would be to give patients with a serious illness a drug that exacerbated their condition. The only way that they could remove the possibility that streptomycin could have harmful effects was by examining patients who would not be exposed to the drug. By helping the investigators to appreciate the limitations of their knowledge about streptomycin therapy, they opened themselves to the idea that streptomycin could be harmful. Investigators discovered that an important new ethical action for them would be to separate their belief about the need for a therapy from their objective knowledge about that therapy's effects. Those who could not would have a difficult time working in the clinical trial era, an observation that is true to this day.

Hill also believed that the high level of efficacy produced by a new therapy could be misleading if that same high level was also seen in the control group. He later demonstrated the importance of a comparison group by revealing that a high success rate for the use of antihistamines to treat the common cold was matched by similar striking findings in a control group [5].

However, acknowledgment of the need for a control group in the tuberculosis study begged the question of which patient should receive the streptomycin as opposed to the control group therapy. As difficult as the fight to include a control group was, the struggle between the clinicians and Hill over therapy allocation would prove to be tougher.

1.5 The Dilemma of Randomization

The random allocation of an experimental intervention is a hallmark of modern experimental design.* The use of random treatment allocations was catapulted to prominence in the mid 1920s by the statistician Ronald Fisher [17,18]. Although Fisher's name is most commonly associated with the use of inference testing in statistics (about which we will have more to say in Chapter Two), he was also one of the pioneers of the use of randomization in research.

Because Fisher worked in agronomy, the first research applications of the random allocation tool were in agriculture. Under Fisher's guidance, new agrarian interventions (e.g., investigational seed formulation or new fertilizer compositions) were allocated randomly to different plots of ground of equal area distributed across the fields. This mix was carefully controlled so that each plot of ground was as

* It is important to distinguish the random allocation of therapy from the random selection of subjects from the population. The random selection of subjects from the population is used in creating the sample, helping to ensure that the sample of patients that is selected for the research is representative of the population from which the sample was selected. The random allocation of therapy occurs after the individual has been selected for the sample. It uses the rules of probability to determine what therapy the patient receives.

likely to receive the new treatment as it was to receive the standard. The resulting patchwork of intervention and control applications helped to ensure that there were no differences between the plots that received the new applications from those that received the standard treatment. Because characteristics of the plots (e.g., proximity to each other, soil moisture and content, insect infestation) did not determine the plot's treatment, these characteristics were removed as possible explicators of the differences in crop yields. This idea of random allocation rapidly took root in agrarian research.

Several years passed before clinical investigators began to explore the possible utility of this procedure for their own work. However, unlike in agrarian research, ethical issues quickly arose in the clinical research arena. It was common for physicians to select the treatment of the research subject. This decision process was simply a natural extension of the habit pattern of physicians in practice who chose the medication for their patients. Therefore, both patients and physicians were comfortable with this historical approach to treatment allocation in clinical research.

Nevertheless, traditional motivations for the therapy allocation contained capricious elements. Inextricably embedded in the decision process were judgments based on the patient's characteristics (e.g., their severity of illness, gender, ethnicity, or financial status). As long as the selection criteria considered characteristics of the patient, it would be impossible to clearly attribute the result seen at the end of the research to the therapy itself.* The random allocation of therapy would solve this problem by creating the environment in which the only difference between patients who receive the intervention and those who did not is the intervention itself, the attribution of effect would be clear [19].

Early efforts at implementing this procedure in clinical research were first attempted in the United States. In 1931, twenty-four individuals who were institutionalized at the Detroit Municipal Tuberculosis Sanatorium were recruited for a study [20]. These cases were individually matched, producing twelve pairs of patients. For each pair, a coin was flipped, and the result of the toss determined which patient of a pair of two would received the active therapy (sanocrysin and sodium-gold thiosulfate injections) versus control group therapy.

Seven years later, 1640 subjects at the University of Minnesota volunteered to receive one of four treatments (three treatments were vaccines, the fourth was a placebo) for the prevention of the common cold. Each student believed that he had received a vaccine when, in fact, the therapy that he received was randomly selected [21]. However, many physicians rebelled against this concept of allowing chance to select the therapy of choice, and the random selection mechanism was prevented from entering the mainstream of clinical research for two decades.

The idea of randomization as a tool re-emerged in Hill's tuberculosis clinical trial fifteen years later. What Hill sought was an allocation mechanism that did not consider personal data, and he believed that the only alternative selection mechanism would be a random one. However, Hill's suggestion that randomization

* This is because those factors that determined therapy allocation would be confused, or confounded with the therapy selection; this confusion makes it difficult to attribute the differences in clinical findings between the control and treatment group to the therapy.

be used in the tuberculosis study ignited a firestorm of debate among its investigators. Physicians could understand the problems generated by poorly planned therapy allocation decisions (e.g., giving the active therapy to only men, and control therapy to only women). However, the notion of making a therapy choice based on the flip of a coin was alien to most, and abhorrent to some.

The motivations for their strong feelings are clear, and resonate to this day. Physicians are trained to be patient oriented. This patient orientation leads us to bring the best of our knowledge, training, experience, and expertise to the patient's bedside. Specifically, when we construct a treatment regimen for a patient, we do it using all of our knowledge about the patient on the one hand, and our expertise with medications. The resultant treatment plan is custom-made for the patient. Woe to the physician who, at the bedside, in front of the patient's family, flips a coin to determine what therapy the patient will receive!

Yet flipping a coin is exactly what randomized therapy is. Hill was obliged to patiently and repeatedly explain to skeptical clinicians what the word "random" really meant. To most clinicians and laymen, then and today, a random process is one that is unplanned, unpredictable, and haphazard. To them, weather could be random, but not a patient's therapy. However, to Hill, random meant a systematic approach in which probability, governed by well-understood laws of chance, would be allowed to prevail. Hill patiently explained that by using chance rather than choice to select the treatment assignment [22], the experiment would provide the independent assessment of a therapy effect, allowing one to "equalize in the two groups the distribution of other characteristics that may be important" [23].

Although Hill had won the fight to include a control group in the tuberculosis study, there is controversy about his success in incorporating the random allocation of therapy. Some suggest that he followed a formal randomization procedure using envelopes completed at a central office that contained each patient's therapy assignment [5]. Others claim that Hill was unsuccessful in persuading the clinicians of the advantages of the random allocation of therapy. These sources argue that the dogged resistance of the physicians to the concept of randomized therapy ultimately led Hill to set the randomized approach aside, replacing it with a strategy of alternating therapy (i.e., the first patient gets active therapy, the second gets control therapy, etc.), a strategy that was more palatable to the investigators [24]. In either case, the trial could only proceed when he avowed to accept a full share of the ethical responsibility for these new trial designs. This willingness on his part was an important reason why clinicians agreed to participate in the studies that Hill designed [5].

Ultimately, the idea of the random allocation of therapy has embedded itself into good clinical trial methodology. However, there continue to be difficulties with its acceptance by some workers, as the following event demonstrates.

> In a randomized, unblinded, multicenter trial designed to compare the effect of different strategies for reducing diastolic blood pressure on the occurrence of strokes, a nurse with established clinical credentials was placed in charge of randomizing patients to either control or active treatment at one clinical cen-

ter.[*] One of the patients recruited into this study was an elderly gentleman. Although the patient met the eligibility criteria for the clinical trial, he suffered from several comorbid, cardiovascular conditions. The nurse accepted him into the program, followed the randomization procedure, and entered him into the control group. During the subsequent follow-up visits, the nurse and patient became friends. Shortly thereafter, the patient experienced a clinical endpoint and subsequently died.

The nurse was genuinely saddened by her friend's death, and gave his demise important consideration. She reviewed her previous decision to follow the randomization scheme that had assigned him to receive the control therapy, now wondering whether she was involved in, if not responsible for, his death. After some reflection, she concluded that her patient should have received more aggressive treatment for his hypertension. Deducing that it was the patient's comorbidities, in combination with the absence of active therapy that killed him, she resolved that clinically ill patients would never receive control group therapy at her center. From that point on, any patient who, in her estimation, had not only hypertension, but suffered from other related conditions (e.g., congestive heart failure, diabetes mellitus, or a prior heart attack) would receive active therapy. If the randomization procedure suggested otherwise, then she would merely alter it in this regard.

The outcome of this decision to use active therapy in the sicker patients at this one center was predictable. This allocation of therapy produced a "canceling out" effect, where the beneficial "positive" effect of the medication was canceled by the "negative" effect of the comorbidities' presence. Because this cancellation did not take place in the control group, a systematic bias was now in place that would underestimate the effect of the active antihypertensive treatment. Undoing the randomization process had confused the effects of the therapy with those of the comorbidities, diluting the effect of the medication on the stroke rate at this center.

Although it is easy to criticize this nurse, careful consideration reveals a deeper, more fundamental issue than the mere inappropriate use of her authority. This nurse's only wish was to deliver the best possible care that she could for her patient. However, she was unable to separate her complete belief in the therapy from her true lack of knowledge of the treatment's effects. Comfortable in her belief, this nurse could not stand idly by while a machine made what, in her view, were inappropriate treatment decisions. Her reaction resonates with physicians and nurses who come into research with a strong practice background.[†]

---

[*] This occurred before the days of computer-generated randomization procedures, that were instituted for, among many reasons, increasing the difficulty of violating the randomization protocol.

[†] The problem is much less common among the new generations of clinical trial methodologists, that is, research investigators and their project managers.

Clinical trial methodologists have effectively and persuasively argued that randomization is necessary in clinical trials [25]. In addition, advances in its implementation have been developed (stratified randomization and adaptive randomizations are but two examples) to more flexibly incorporate its advantages into clinical studies. Nevertheless, many of the clinicians whose patients are selected for these studies continue to struggle to understand the necessity of a procedure that appears to be the antithesis of the good practice of medicine. Nevertheless, randomization is the only currently available procedure ensuring independence between a patient's characteristics and their research therapy allocation.

1.6 Blinding

The final adaptation that Bradford Hill introduced into the streptomycin study was a blinding mechanism that masked knowledge of the therapy assignment. In his tuberculosis study, patients were not told what treatment they were receiving. In fact, these patients were not even told that they were participants in a study! [5]* Although this last adaptation is unacceptable in our contemporary research environment, the utility of blinding is uncontested.

Blinding in a clinical study protects the study from influences that can distort the size of the treatment effect. In the previous section, we stated that the motivation for the use of the random allocation of therapy in a clinical trial is to ensure that the only difference between subjects who receive the intervention to be studied and those who do not is the therapy itself. Thus, at the time of the therapy assignment (commonly referred to as the baseline), the distribution of all patient characteristics (e.g., demographics, lifestyle, previous medical history, and physical examination findings) is the same between the two groups; the two groups of patients are equivalent except for the therapy exposure.

Unfortunately, beginning a clinical trial with equivalent patient groups does not guarantee that the trial will end with this equivalence property intact. If the investigators are to be assured that any difference that is seen between the active group and the control group at the end of the trial can be ascribed to the randomly allocated therapy, the two groups of patients must not only have equivalent characteristics at the baseline; the patients must also have equivalent experiences during the study (e.g., equal compliance with the assigned therapy) excepting the effects of the intervention. Ensuring this equivalent post-randomization experience is complicated when the patient and/or the physician knowing the identity of the medication that the patient is taking. Blinding is the collection of procedures that restrict knowledge of the treatment identity. Their implementation increases the likelihood that a patient's post-randomization experience will reflect the effect of the intervention and nothing else.

For example, if a patient knows that she is on placebo therapy, she may believe that her condition is more likely to deteriorate than to improve. This will lead to actions that are motivated, not by the action of the study medicine to which

* This was not the first time a clinical study was blinded. In the Detroit Sanatorium study discussed previously, patients were not told which therapy they were placed on (sham subcutaneous injections of distilled water served as the placebo therapy in that experiment).

she is assigned, but by her perceptions of what that therapy is. This patient may adjust other medications that she is taking based on her belief in the ineffectiveness of the study medication. This same perception can lead her to provide relatively negative quality of life reports and self-assessments to the clinical investigator. If this conviction takes root in the majority of patients who are assigned to the control group, the investigators might conclude that the placebo experience is less satisfactory than it actually is. Alternatively, patients who know that they were assigned to active therapy may be inclined to believe the therapy is helping them. This belief has its own invigorating effect.

The influence of these belief systems is strong and, if left unchecked, will blur the investigator's view of the therapy's true effect. In order to counterbalance the influence of these constructs, investigators instituted single-blind trials in which patients were not informed of their therapy assignment. In these single-blind studies, patients do not know whether they are on active therapy or placebo therapy.

It is important to note that the single-blind approach does not suppress the belief system; it simply distributes it equally across the two groups of patients. This distribution spreads the effect throughout the entire cohort, rather than concentrate it in one group, thereby removing it as a source of bias.*

Physician knowledge of the medication that the patient is taking can also skew the objective evaluation of the effect of the therapy. Physicians commonly agree to be a participant in these studies because they have feelings (sometimes strong feelings) about the effect of the compound that is being studied. These strong feelings can influence the way a physician treats a patient during the course of the study, leading them to (1) insist that the patient be compliant with the medication, and (2) express to the patient the importance of returning for all of the scheduled follow-up visits that the clinical trial requires.

In addition, doctors may choose to use other concomitant medications much more aggressively in patients who do not receive active therapy. At the end of the follow-up period, these same physicians may be more diligent in seeking out adverse outcomes from patients who are randomized to the treatment group of the trial that the investigator believes is ineffective and/or produces more side-effects. Each of these maneuvers can adversely affect the assessment of the therapy's influence. In order to distribute these effects randomly among the physicians who treat patients in the study, these investigators are blocked from knowing the medication that their patient is taking.

Studies in which neither the physician nor the patient knows the effect of the therapy are known as double-blind trials. The 1938 University of Minnesota study† was not only one of the first randomized studies, but was also one of the first investigations in which blinding was implemented. In that circumstance, neither the randomized students nor the physicians knew which treatment any individual student received [21].

---

* Bias is a systematic influence that distorts the treatment effect measure. For example, in an unblinded trial, the negative perceptions of patients who know that they are in the control group affect only the control group, thus providing a misleading measure of the effect of the therapy.

† This study was discussed on page 8.

Double-blind studies can be difficult to sustain because of the known effects of the medication that are not mimicked by placebo therapy. A fine example of (1) the need to maintain a double-blind in a clinical trial and (2) the lengths to which investigators must go to preserve the double-blind property of a clinical trial is the evaluation of the use of arthroscopic surgery as a tool for relieving the pain and disability associated with osteoarthritis of the knee in which sham surgery was implemented [26].

However, as useful as the blinding mechanism is, double-blind trials can put the treating physician in a difficult situation. For treating physicians, the fact that the therapy that was selected for a patient is based on chance and not patient-specific knowledge is a difficult enough concept. Their practice difficulties are compounded by the ignorance of the therapy that was ultimately selected for the patient. Physicians intervene to save lives and minimize morbidity by swift and decisive action. This decisive action is based on firm knowledge about the patient; a key component of that knowledge is the patient's history, including his record of medication use. Physicians commonly don't know how they are going to deliver safe and effective care for their patients within the strictures of these binding limitations.

Consider an example from the CARE (Cholesterol and Recurrent Events) Trial [27]. This study was designed to examine the effect of the HMG-CoA reductase inhibitor pravastatin on the occurrence of clinical endpoints in patients with normal levels of low-density lipoprotein (LDL) cholesterol. When the study was carried out in the early 1990s "normal" LDL-cholesterol levels were considered to be in the 115–175 mg/dl range.

CARE patients were randomized to receive pravastatin or placebo. It was accepted at the time that pravastatin would reduce LDL-cholesterol levels by approximately 25%. The study was designed to test whether this effect on LDL-cholesterol levels would translate into a reduction in the primary endpoint of fatal and nonfatal myocardial infarction (MI). Patients would be followed for five years.

Because CARE was a double-blind trial, the clinical investigators did not know the therapy assignment for any of their patients. However, because pravastatin had such a profound effect on lipid levels, an examination of the patient's LDL-cholesterol levels in her chart would reveal the patient's assigned therapy. Therefore, not only were the physician and patient blinded to therapy assignment; they were also blinded to lipid levels. Physicians who followed these patients were told to send blood drawn for lipid levels to a central certified laboratory where lipid assays would be run.

The CARE clinical and data coordinating centers worked together to follow these lipid measures. If a patient's LDL cholesterol level was too high, the clinical center was contacted, and the physician investigated the circumstances, encouraging the patient to resume study medication if the patient stopped taking the assigned medication, or to add adjunct therapy. Because the necessity of the maneuver for one of his patients could unblind the physician, sham instructions were also provided for a patient in the opposite treatment group at the same clinical center. However, this was not a perfect mechanism as the following example shows.

"You are going to lose a patient from your trial, and worse, I am going to lose a patient from my practice because of this study," a CARE clinical investigator raged at the coordinating center over the phone. This physician had enrolled a patient into the study and had been following this quiet gentleman carefully. All blood work drawn for this patient was, per protocol, sent to the central laboratory. Over the course of the follow-up visits, the patient would occasionally ask how his cholesterol levels were doing, to which the physician replied that the cholesterol levels were fine. After all, if the cholesterol levels were higher than 175 mg/dl, then the doctor would have heard from the coordinating center. Thus, the clinician felt comfortable in reassuring the patient who was always gratified by this news.

However, today's visit was different as both the patient and his wife, were present. The wife, clearly angry, demanded that her husband remove himself from the study. She would not stand by while her husband's care and treatment were neglected! Her husband, normally quiet, was also visibly agitated.

When the clinical investigator asked why they were so upset, the patient's wife began by explaining that she and her husband had gone to a mall to do some shopping. While there, she had noticed that the mall management had arranged for shoppers to have their cholesterol levels screened. When she suggested to her husband that he be checked, he declined, saying that "The CARE physicians had been monitoring my cholesterol." However, she insisted because, "You've already had one heart attack. What's the harm in checking?"

She was shocked a week later to learn that her husband's LDL cholesterol level was over 225 mg/dl, "in the dangerous range". She confronted her husband with the results, and now here they were in the clinical investigator's office, confronting him.

Discussions commenced between the coordinating center and the clinic physician in an attempt to resolve this difficulty. They concluded that the best action to take would be to confirm the elevated LDL-cholesterol level. Blood was drawn although the patient was in his office, and the certified central CARE laboratory agreed to process the specimen quickly. The patient, his wife, and the clinical center were gratified when the central laboratory's result revealed that the LDL-cholesterol level was much lower than the mall's result. The best explanation for the disparity was the difference in the quality of the two laboratories.

With no knowledge of the patient's LDL-cholesterol, the clinical investigator was at a disadvantage in working with his patients, and was unprepared for the third-party LDL-cholesterol assay.

1.7 Hill's Results

After Hill's steadfast and persistent efforts, the tuberculosis study was started with (1) alternating therapy selection, and (2) blinding. The total study consisted of 107 patients, all of whom were admitted to the hospital in 1947. The study treated and followed patients for one year and produced clear results (Table 1.1).

Table 1.1. Results of Tuberculosis Clinical Trial

Follow-Up Total patients	Streptomycin Group $n = 55$		Control Group $n = 52$	
	Patients	Percent	Patients	Percent
6-month mortality	4	7	14	27
12-month mortality	12	22	24	46

Both six-month and twelve-month crude mortality rates were reduced in the streptomycin group. The difference at twelve months was statistically significant [5].

The study was accepted as a clear success. The simultaneous introduction of three devices (a control group, external therapy selection mechanism, and blinding) in the same experiment improved the ability of the scientists to draw clear and unambiguous conclusions about the effects of streptomycin. Much was made of the fact that very little statistical analysis was required [28]. However, the same tools that provided such clarity also generated a new ethical concern, from which arose the need to monitor clinical research in a new and imaginative way.

1.8 The Monitoring Rationale

The effective use of (1) a control group, (2) randomized therapy allocation, and (3) double-blind masking procedures can produce unambiguous clinical trial results. However, as is commonly the case in the evolution of science, each step out of an old problem is a step into a new one. The new problem in this circumstance was the occurrence of unanticipated effects in blinded clinical experiments.

Investigators designing a trial give important consideration to many factors. Two of these are (1) the primary analysis of the study, and (2) the duration of time that patients will be followed. The primary analysis is that specific quantitative evaluation (e.g., mortality effect) that will determine whether the study is positive. The duration of follow-up is the anticipated time that the intervention will need in order to produce its measurable effect on the primary endpoint in patients. The duration of follow-up depends on the therapy and the endpoint. For example, if one wishes to determine the effect of angiotensin converting enzyme inhibitor (ACE-i) therapy on the blood pressure of patients with mild diastolic hypertension, one need only follow patients for a few weeks. However, if one wishes to observe the effect of these same agents on stroke rates, one must follow these same patients for years.*

Unfortunately, good planning does not always produce accurate predictions. Because the question of the effect of therapy is an open one, it is difficult for investigators to know precisely how long it will take the therapy to reveal its effect. Also, because the research sample is just one of many possible samples from the population, the play of chance can produce an aberrant therapy effect (either harmful or helpful) in the sample that does not represent the true therapy effect in the population. These two influences can combine to provide surprising results at unanticipated durations of follow-up. Most disconcertingly, these effects can occur quite early during the conduct of a clinical trial, demanding immediate action on the part of the investigators. However, if the investigators are blinded to the therapy assignments, they cannot observe and assess these early effects.

An early reflection of this concern was the desire to identify a procedure by which a study could come to an early conclusion without the need to either randomize an unnecessarily large number of patients, or to expose patients senselessly to a therapy that was proving to be inferior. This desire was satisfied by the adaptation of the tool of sequential testing.

1.9 Early Monitoring Tools: Sequential Trials

Like many aspects of clinical trials, the concept of sequential testing is simple. Patients are recruited into a clinical trial two at a time. One member of the pair is randomly chosen to receive active therapy; the second receives the control treatment. After a short period of time, the response to therapy of each of the two patients is assessed. A second pair of patients is then recruited, following the same procedures as were followed for the first. The information of this second pair is added to that of the first pair to compute a cumulative effect of the therapy. A decision is now made. If this cumulative analysis demonstrates a clear benefit of the therapy, the trial is stopped in favor of the active therapy. If the trial demonstrates a clear disadvantage associated with the active therapy, the trial is stopped in favor of the control group. If neither benefit nor hazard is demonstrated, the trial is permitted to continue, and another pair of patients is recruited. Thus, the trial is analyzed sequentially after each pair of patients is treated and evaluated. Trials that are designed to function in this way are sequential clinical trials.

The idea for sequential testing was first generated in industry, where products have to be inspected for defects [29]. The procedure was adapted to clinical trial use by Abraham Wald, who attributed his ideas to discussions at the Statistical Research Group in Columbia and the Bureau of Ordinance at the Navy Department in 1943 [30]. Wald created the Sequential Probability Ratio Test (SPRT) to assess the early effect of therapy in these studies.† Peter Armitage outlined the basic modern principles of sequential clinical trials, writing what has become the seminal text

---

* Because blood pressure changes can be measured in each patient, enough information is available that permits an early assessment of the effect of the therapy. Strokes, on the other hand, are relatively uncommon. Therefore, patients must be followed longer in order for enough strokes to occur in study participants.

† Wald died shortly thereafter in 1950 during an airplane accident.

for the implementation of this methodology [31]. According to Armitage, one of the most important advantages of this approach is the ethical consideration that obviates the unnecessary use of any inferior treatments. In addition, less data was generally required to come to a conclusion about the effect of therapy. A final advantage, as pointed out by Bross, is that, because the computational burden was not immense, the analysis could be carried out in the absence of a statistician [32]!

However, although this innovation marked the first development of a systematic procedure in which the effect of therapy could be assessed in an ongoing manner, the sequential monitoring design has important restrictions. The requirement that the monitoring evaluation be executed for each pair of recruited patients demands that there be a rapid determination of efficacy [33]. This restriction precludes its use in studies where the endpoints may not occur until months or years have elapsed.

1.10 Data Monitoring Committees

Clinical investigators had to solve the issue of identifying the unanticipated early appearance of a therapeutic benefit or hazard that was not expected to emerge for months or years without unblinding the investigators of the study. This produced the need for Data and Safety Monitoring Boards (DSMBs), now known as Data Monitoring Committees DMCs [34].

The concept of a monitoring committee arose in the 1950s, emerging from the ferment of ideas produced from the careful study of early clinical trials following Hill's tuberculosis study. Evolving informally for approximately a decade, the popularity of these groups increased in the clinical trial community. However, their structure and function solidified when the National Institutes of Health undertook a review of the process in the 1960s. The resulting report of Bernard Greenberg served as a blueprint for the organization and oversight of large, multicenter clinical trials.

A DMC is a relatively small collection of august scientists who have particular expertise in the clinical question addressed by the trial. This distinguished group of scientists commonly includes clinicians, clinical researchers, methodologists (e.g., biostatisticians or epidemiologists), and sometimes, an ethicist. The central charge of this group is to review all of the data in its blinded and unblinded form* and to determine if either an early therapeutic triumph or early therapeutic catastrophe has occurred.

The mandate of the DMC has expanded in recent years to review not just the effect of therapy but to examine other barometers of the trial's status as well. It is not uncommon now for DMCs to review the progress of patient recruitment, to be notified of protocol violations, and to ensure that the statistical and epidemiological assumptions that underlie the sample size computation are correct. This group then makes recommendations to the investigators and the sponsors of the study concerning the clinical trial's status. It is the responsibility of this group to

---

* Sometimes even the DMC is blinded to the identity of the therapy effects. In this "triple-blind study," the DMC sees the results categorized not as active versus control group findings but, instead, as "Group A" and "Group B."

ensure that the trial is executed according to its protocol and to be on the alert for early signs of therapeutic benefit or harm.

The Greenberg report called for a mechanism that would permit the early termination of a clinical trial when unanticipated early information obviated the trial's continuation [34]. However, exactly how the DMC was to make a determination of early benefit or harm was an open statistical question. The dataset that is collected by a well-designed and well-executed clinical trial is a random sample from a much larger population of patients. That sample very likely contains a signal that identifies the true nature of the relationship between the intervention and the disease in the population from which the sample was obtained. However, samples also contain, just through their random aggregation of subjects, false leads. This background noise can distort and, in some cases, overwhelm the efficacy signal.

It can be difficult to correctly discern this signal at the conclusion of the research effort, when all patients in the study have completed their follow-up and have had their data collected and analyzed. The task that Greenberg placed before the DMC was to identify this signal when only a fraction of the trial's total time has elapsed, and only a few patients have data to be analyzed. This was a hazardous undertaking. Bradford Hill [35] wrote a prescient comment about the problems that would face such a committee. The committee, he wrote,

> "...has to walk a wobbly rope, balancing between panicking over something that is suggestive but may turn out to be unimportant and doing nothing because the case is unproven and yet may turn out to be true so that harm is meanwhile done."

1.11 Gaining Experiences in Monitoring: CDP

One of the first randomized clinical trials that demonstrated the difficulties presented by the interim monitoring of clinical trials was the Coronary Drug Project (CDP). Executed in the 1970s, this trial was designed to assess the effectiveness of different strategies in reducing the risk of atherosclerotic disease in men.*

CDP was one of the largest projects of its time, randomizing 8341 patients from 55 clinical centers. Each patient was randomized to receive either placebo or one of the following five therapies; (1) low-dose estrogen, (2) high-dose estrogen, (3) clofibrate, (4) dextrothyroxine, or (5) nicotinic acid. Patients were followed for a minimum of five years.

Early interim evaluations of the effect of therapy were carried out by the DMC of the study. It was clear to its members that (1) both of the estrogen doses

---

* A reasonable question to ask is why the CDP investigators chose to exclude women. Although cultural issues during this period cannot be excluded, one motivation derives from the sample size consideration. Because the trial's principal evaluation would be based on a comparison of the number of endpoints in each of the study's treatment groups, enough of these endpoints must occur for a precise assessment. In general, the rarer the endpoint, the greater the number of patients required to collect a sufficient number of them. Because, in general, atherosclerotic cumulative morbidity and mortality rates are greater in men than in women, the sample size of the trial and the consequent resources required to execute the trial could be reduced by focusing only on men.

and 2) the dextrothyroxine treatment arm were ineffective: these study treatments were discontinued. However, continued interim evaluations of the effect of therapy suggested that there might be a harmful effect of clofibrate on the cumulative total mortality rate. In fact, during the first 40% of the study, the trend was for excess mortality in this treatment arm.

It was exactly this type of trend that the DMC was commissioned to detect and to which it was charged to respond. However, the committee members wisely recognized that such early results could be influenced by two factors. The first was that the excess early deaths that occurred in the clofibrate treatment arm were occurring simply because the vicissitudes of chance had produced for the investigators a pattern of death that was not truly reflective of the effect in the population. The second is that the estimates of high early death rates were based on a relatively small number of deaths in a small cohort with a short follow-up period. If the number of patients recruited into the trial increased, and the length of follow-up time increased, then better, more precise estimates would be obtained.

This decision not to stop, but to continue to follow the clofibrate patients to the study's end was a controversial decision by the DMC. If mistaken, this action could have slowed the acceptance of these large-scale, expensive studies by the medical community. However, by the time the study ended, the harmful cumulative mortality trend had disappeared, and clofibrate was found to be equivalent to placebo therapy [34].

The DMCs decision to continue the study was sound. However, one of the surprising findings to observers of this study was that an early deleterious trend for clofibrate could be so misleading about the final effect of that therapy. Specifically, a therapy that proved in the end to have only a placebo effect could, just through the play of chance, produce a trend that "masqueraded" as a harmful one, Apparently, therapies could produce very misleading short-term trends about their long-term effects.

This was a new complication. It was well known that the analysis of the final results of clinical trials could be a perplexing process. However, through the joint albeit complex management of sound methodology, *p*-values, magnitudes of effect, and confidence intervals, the research community had gained some intuition and comfort with comprehending clinical trial results. Now, the successful interim monitoring of these experiments would require a entirely new level of intuition.

Very few workers at the time anticipated that a useful key to this ethical and clinical problem would be held by the gentle movement of pollen grain.

1.12 Ceaseless Agitation

Robert Brown's independent thought would either make him or break him as a scientist. Born in Scotland in 1773, he was the son of the fiercely independent Scottish Episcopalian minister, James Brown, who was himself widely known for his strong intellect [36]. Educated at the Marischal College in Aberdeen, Robert went on to study medicine at the University of Edinburgh. After graduation, he focused on his developing love for botany, leaving time only for the study of German grammar in the morning and seeing patients from 1:00 PM to 3:00 PM in the afternoon. His botanical appreciation was fostered by a sustaining relationship with Sir Joseph Banks, an eminent botanist.

Brown's particular focus was on the use of the microscope to describe the anatomy of the reproductive systems of plants. During these intense sessions of careful still observation, Brown could not help but notice the incessant agitation of the minute pollen particles in an aqueous suspension, vibrating with an unexplained force. The movement was always apparent, but not uniform. This activity was complex, generating movement in three dimensions. However, it one were to consider just one-dimensional movement (i.e., up or down) and plot this movement over time, the motion appeared to be haphazard with no discernable purpose (Figure 1.1).

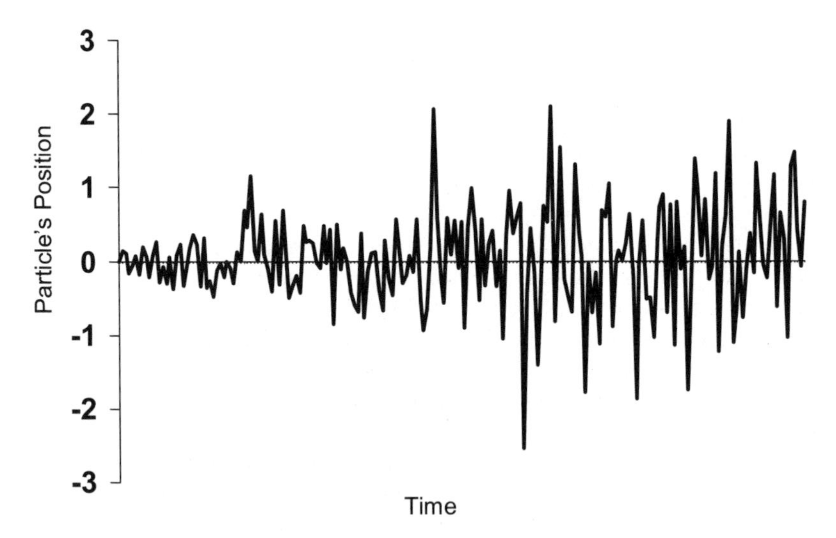

Figure 1.1. Simulation of the one-dimensional movement (i.e., either up or down) over time that is the hallmark of Brownian motion.

Disciplining himself, Brown entered into a period of intense study and careful explanation of these findings. After his well-considered research revealed that the movement could not be explained by currents in the fluid in which the pollen grains were suspended, nor by gradual evaporation of this fluid, he revealed his findings to others.

The tepid response to Robert Brown's revelations quickly disabused him of the idea that he had discovered something of value. Careful microscopists had identified this type of movement before,* and it was felt that the phenomenon was

* The phenomenon was first observed by Jan Ingenhouz in 1785, according to sources used by Eric Weisstein's *World of Physics*, which can be found on the Internet at http://scieneworld.wolfram.com/physics/BrownianMotion.html.

of no real importance. The current, accepted dogma was that this motion was a property of the pollen grain, a particle that, in the early 19th century, was assumed to have life all of its own [37]. However, Brown's intellect would not permit him to accept this explanation without evidence, and his independent spirit drove him on to identify a more exact explanation for this movement. His demonstrating of this phenomenon within pollen grains that had been preserved for years in an alcohol-based solution voided the argument that this vibration was a force of life. However, Brown's revelation that even rocks, when broken into sufficiently small particles, exhibited this same motion, led his adversaries to concede that the movement was neither a property of life, nor of organic substances. The movement was a physical property.

His willingness to commit his combination of independent spirit and intellectual prowess produced a rigorous counterargument to these accepted explications. His exertions earned the property of matter that he studied the sobriquet, "Brownian movement", that was later renamed Brownian motion. Brown died in 1858.*

1.13 Early Explanations for Brownian Motion

Although Robert Brown revealed that Brownian motion was a physical property of small particles, a definitive explanation for its cause was not within his grasp. Debate ensued through the middle to the end of the 19th century for the root cause of this movement.

By the 1860s, the persistence of the phenomenon and the absence of a unifying explanation drew the attention of theoretical physicists. The characteristics of the motion were quite simple and remarkable. It was agreed that, when a particle was exhibiting Brownian motion,

1. It was equally likely to move in any direction.
2. Its future motion was unrelated to its past motion.
3. Its motion never stopped.

These central properties of Brownian motion have been adopted by clinical trial methodologists to describe the movement of a function of a test statistic during the course of a clinical trial. However, at the time, this perpetual unpredictable (or to use the language of the time, irregular) movement generated excitement in the field of physics.

Detailed investigations determined that the combination of a small particle size and a low fluid viscosity produced faster motion [38]. This concept, when considered with an earlier observation that higher temperatures led to more rapid Brownian motion, suggested that a possible explanation lay in the molecular motion that was occurring in the surrounding liquid. However, ultimately, a comprehensive and satisfactory explanation of this 19th century phenomenon had to wait for the

* He died on June 10, 1858, just a week before Charles Darwin received Wallace's paper on the theory of "survival of the fittest." In fact, Brown's death freed up a date at the Linnean Society at which Darwin could present his own controversial and order-shattering lecture on the theory of evolution.

mind of the most eminent scientist of the 20th century, Albert Einstein. By offering a unifying explanation for Brownian motion, Einstein quite inadvertently provided a tool for monitoring clinical research that would lay dormant for 60 years.

1.14 Enter the Mathematicians

In his doctoral dissertation submitted in 1905 to the University of Zurich, Albert Einstein developed a statistical molecular theory of liquids [39]. In a separate contemporaneous paper, he applied this theory to identify the effect of molecules on a larger mass. He could not have imagined the implications of this work for monitoring clinical research.

The idea that a liquid or gas is made up of molecules that are always moving and colliding with each other was a core component of the kinetic theory of matter. Without performing any experiments, Einstein reasoned that, if the theory of the existence of atoms was correct, then atoms in a liquid, speeding along on their "infinitely many" paths, must have collisions with each other and with larger particles. The effect on these larger particles would be movement, but what kind of movement would that be?

An exact solution to this problem was theoretically available through the application of Newton's well-known equations of motion and force. However, this solution would have to take into account each of the billions of moving molecules and collisions each instant. No such practical solution to this theoretical problem was available. Einstein took the novel and innovative perspective of solving the problem statistically (i.e., in the aggregate) rather than providing a precise solution for the result of every collision [40].

Einstein thought that the resulting effect of these collisions on larger particles would be movement. This motion was due to the difference in pressure produced by the collisions on one side of the particle from the collisions on the other side. Because this "pressure differential" was random, changing from one moment to the next as the collisions changed randomly, the movement of the particle would be random.

Einstein then took the next step to quantitate this movement. He focused on the one-dimensional movement of the particle (i.e., a limited form of motion in which the particle could only move in one dimension, either up or down).* Specifically, over a period of time, the particle would tend to drift from its starting point, moving randomly up then down over time. Kinetic theory made it possible to compute the probability of a particle P moving a certain distance x from its starting point in relation to the elapsed time t and the viscosity of the fluid D. Einstein also used the kinetic theory of gases to derive the diffusion constant for such motion in terms of fundamental parameters of the particles and liquids [41].† On the basis of

*This is called one-dimensional Brownian motion. Molecular particles in reality exhibit three-dimensional Brownian motion, simultaneously oscillating up and down, side to side, and in and out. Multidimensional Brownian motion is a more complicated topic and is beyond the scope of this discussion.

† This work was used by Perrin as a basis for the computation of Avogadro's number.

this work, Einstein deduced that $P = \frac{e^{-x^2/4Dt}}{2\sqrt{\pi Dt}}$. But with just a little algebra, the quantity P may be rewritten as

$$P = \frac{e^{-x^2/4Dt}}{2\sqrt{\pi Dt}} = \frac{e^{-x^2/2(2Dt)}}{\sqrt{2\pi}\sqrt{2Dt}} = \frac{e^{-x^2/2\sigma^2}}{\sqrt{2\pi\sigma^2}} \,. \tag{1.1}$$

By letting $\sigma^2 = 2Dt$, we recognize the quantity on the right-hand side of expression (1.1) as the formula for finding probabilities under the standard normal or "bell-shaped" distribution. This curve typically arises when the variable of interest is the sum of many independent forces. Einstein's derivation was a natural and satisfying "fit" to the idea that the microscopic particle's motion is the sum of random pushes in different directions.

However, perhaps the most remarkable observation about Einstein's particle motion work was that Einstein knew nothing of the observations of Robert Brown [38]! Einstein did not know that the process he had just described quantitatively had in fact been observed a century earlier. This independent evaluation and identification of Brownian motion cemented the veracity of the kinetic theory of gasses in the minds of physicists.*

1.15 Trajectory Mathematics

Einstein developed the notion that a statistical treatment of complex systems could provide useful insight into the behavior of complex systems (like the trillions of molecular bombardments that a microscopic pollen granule sustains in suspension). However, it was difficult to do anything of practical importance with this result, primarily because the mathematics of probability had not yet been sufficiently developed. However, this last giant step that moved this field towards the application of clinical science was undertaken by a mathematical prodigy, Norbert Wiener.

The first problem to which Norbert Weiner turned his attention was the mathematics of Brownian motion.† Much had happened in probability because of Einstein's 1905 publication fifteen years earlier, and Wiener was intrigued by the new possibilities. Was is possible to compute the probability that a particle would

---

*The French physicist Jean-Baptiste Perrin was successful in verifying Einstein's analysis, and for this work was awarded the Nobel Prize for Physics in 1926.

† The story of Norbert Wiener is not just the demonstration of the power and idiosyncrasies of a mathematical personality. It is a story of remarkable parental persistence and optimism. Norbert Wiener's father responded to his son's poor academic performance by personally removing him from school and tutoring his son himself. The result was miraculous. Six years after he professed a crippling and limiting weakness in mathematics, Norbert Weiner was admitted to Tufts College at the age of eleven. He completed his undergraduate studies when he was fourteen, upon which time he was admitted to Harvard graduate school. Pursuing training first in zoology, and then mathematics he completed his PhD at 18. Shortly thereafter, he became a mathematics instructor at MIT, where he remained for his entire career. He was 24 years old.

follow a pre-specified trajectory? Weiner recognized that it would be impossible to predict the precise path that a microscopic particle being bombarded by trillions of molecules every moment along the way would follow. The number of equations that would need to be solved would be far beyond the calculating ability available. However, he reasoned, it might be possible to determine the likelihood of certain trajectories or paths; to compute, for example, the probability that a particle, after a time period of one second, has moved at least one unit up from its starting point There were many trajectories the particle could follow. A fraction of those would meet this criteria. How likely was this fraction?

Many eminent mathematicians attempted to solve this problem, attempting to link Einstein's theoretical work to this practical application.* Wiener actually did it, by creating (1) a new concept, the stochastic process, and (2) an innovative measure that still bears his name—Wiener measure. Wiener measure permits the accurate computation of these excursion probabilities that are critical to modern statistical monitoring procedures in clinical research. Essentially, clinical trial methodologists today model the movement of a test statistic much as Einstein and Wiener modeled the movement of particles, using the normal distribution.

Wiener himself was somewhat of an eccentric. His speech, like his writings were difficult to understand. All too often, Wiener was unable to resist the temptation to say everything that came into his mind. He could not separate mathematics from its implications, nor its implications from his personal experiences. It was as though in Wiener's mind, the person he was addressing instantaneously changed from a layman, to a mathematician, and then to Wiener himself [42].

It was the first formal demonstration that probability theory could be applied to events that occur randomly in time, and it created the new study of stochastic processes, This field was essentially a creation of Norbert Wiener. The mathematical tools were now available to compute probabilities of trajectories.

1.16 Momentum Builds for Monitoring Procedures

As clinical trials grew in scope and complexity in the 1960s and the role of the DMC crystallized, the need for formalized monitoring procedures accelerated. This led to the development of statistical monitoring procedures that provided a solid structure to guide the DMCs evaluation process. Heybittle [43] and Peto forged important new tools that typified early attempts to provide DMCs with rigorous quantitative advice on terminating a clinical research effort prematurely.

A newer class of procedure was disseminated in the 1970s. This body of calculations, termed group sequential procedures, was the first successful attempt to apply the underlying mathematics used to describe Brownian motion to the development of statistical monitoring procedures in clinical research. They are the basis of the most commonly used clinical trial monitoring rules, and are the computations that produce the graphic depiction of monitoring procedures most familiar to investigators. These figures commonly divide a graph into regions of trial termination or trial continuation (Figure 1.2).

* Including the likes of Borel, Lebesque, Lévy, Banach, Fréchet, and even the preeminent mathematician, A N Kolmogorov.

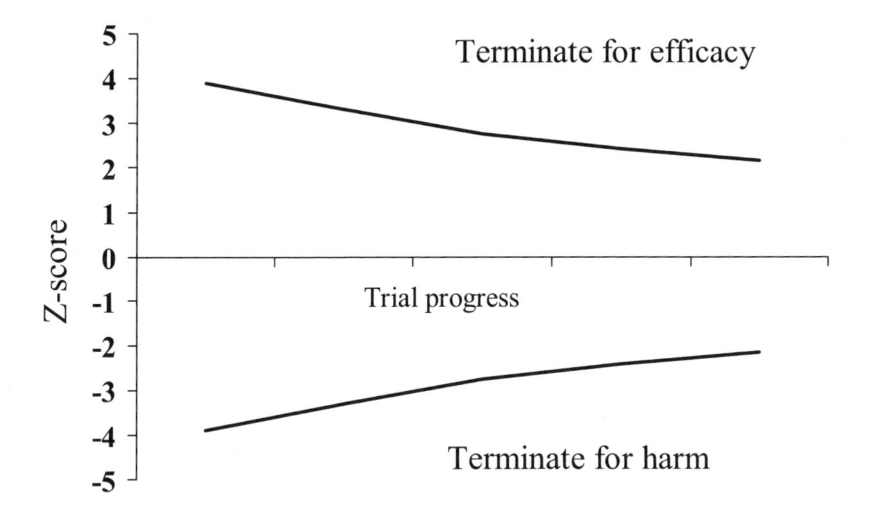

Figure 1.2. Example of Lan–DeMets monitoring boundaries for efficacy and harm in a clinical trial.

Pocock [44] and the work of O'Brien and Fleming in 1979 [45] were the first products of this new area of investigation. Taking its name from its authors, the O'Brien–Fleming procedure is the progenitor of many of the modern monitoring procedures used in clinical trials. Important and useful advances and adaptations of this approach were developed by Lan and DeMets [46,47,48]. Whitehead [49, 50] produced another perspective on these issues. The popular innovation of stochastic curtailment was introduced in the 1980s [51,52,53], and the notion of conditional power was introduced by Halperin et al. [54].

At this point, the vocabulary of clinical trial methodology began to swell with statistical neologisms. One new phrase was group sequential boundaries, or the curves appearing on the test statistic versus time graph that separates one decision choice (e.g., to terminate the trial for harm), from another (e.g., continue the study) (as in Figure 1.2). Group sequential procedures are the body of assumptions and statistical calculations that produce these boundaries. The concepts of stopping for "benefit", "harm", and "futility" entered the clinical research jargon during the early 1980s.

1.17 Bayesian Considerations

Alternative procedures for monitoring clinical studies have considered Bayesian methods. The Bayesian approach is more than the introduction of new statistical methodology. It represents a different statistical paradigm that formally incorporates (1) the use of prior information, and (2) the construction of a loss function that helps to gauge the effect of different decisions on the investigators. The link between Bayesian tools and statistical monitoring of clinical research was explored in the 1980s [55,56,57]. This philosophy has generated discussion [58] and the development of additional methodology [59,60]. Its use as a monitoring procedure in clinical research will be discussed in Chapter Eight.

1.18 "Then they'll get it right..."

Born June 18, 1981, Jesse was a real character in a lot of ways, so much so that his parents considered giving him the name "Jesse James" Geisinger.* Three years later his rambunctious and normal infancy was shattered by the emergence of wild and unpredictable behavior. He became erratic, and his speech patterns rose to levels of harshness and belligerence. Stunned by the sudden appearance of this bizarre behavior, his family struggled with possible explanations. Some thought Jesse was possessed. Others feared that he, like other family members before him, had schizophrenia. However, a trip to the doctor and an adjustment in his diet returned Jesse to normal. His parents could not understand the dramatic beneficial effect of protein reduction in his diet, but they were thankful nevertheless. They returned to their state of twitchy alertness that all parents of active children come to accept.

Three weeks later, Jesse plopped himself in front of the television, went to sleep, and would not wake up. Rushed to the emergency room by his frantic parents, they were told that their son was in a stage 1 coma. Eleven days later, Jesse Geisinger was released from the hospital, three years of age, alert, carrying with him the new diagnosis ornithine transcarbamylase deficiency syndrome (OTC), a rare metabolic disorder.

The doctors thought that Jesse's form of the disorder was mild and could be controlled by medication and a protein-sparing diet. However, as Jesse grew, he continued to struggle with his diet and medications, resulting in metabolic derangements that in turn produced repeated hospitalizations. After discussions among Jesse, his parents, and the physicians at the University of Pennsylvania, Jesse was admitted to a clinical trial in which a viral vector would be used to alter Jesse's defective DNA.

On a Monday, Jesse was infused. That night he had a fever. Tuesday he was jaundiced and disoriented. Within a few days, Jesse Geisinger died.

The resulting investigation revealed many ethical problems with the treatment of Jesse Geisinger, including lax oversight of the research effort and conflict of interest. From the perspective of Jesse's parents

> There is so much more to Jesse's story. I can't help but believe that they will kill this with time and money, as they always seem able to do. Who is "they"? They are heartless and soulless industry and their lobbying efforts; they are the politi-

---

* This brief account of the life and death of Jesse Geisinger is taken from http://www.sskrplaw.com/gene/jessieintent.html.

> cians and bureaucrats more interested in placating industry than in protecting the people, they are doctors so blinded in their quest for recognition that they can't even see the dangers anymore. Let them apply Jesse's intent to their efforts, and then they'll get it right.

In 1998, after this and other ethical lapses in the conduct of clinical trials, the Office of the Inspector General of the Department of Health and Human Services put forward recommendations to help ensure the safety of clinical trial participants that had important implications for monitoring clinical research. Most of these recommendations focused on the conduct of institutional review boards (IRBs) and DMCs. In response, the Federal Food and Drug Administration (FDA) initiated discussion on the appropriate conduct and oversight of clinical research. The resulting document, entitled Guidance for Clinical Trial Sponsors on the Establishment and Operation of Clinical Trial Data Monitoring Committees [61] provides preliminary statements as to the need and constitution of DMCs.

This guidance focuses on the needs, composition, and operation of the DMC in contemporary clinical research. In addition, this report reinforced the rationale for the use of monitoring procedures in clinical research. The need to control overall false positive error rates during the sequential monitoring procedure, and the requirement to end a clinical experiment early in the face of clear demonstration of benefit, harm, or futility were clearly acknowledged. As a result of this investigation, the use of statistical monitoring devices dramatically increased.

Therefore, at the turn of the century, the well-motivated, mathematically challenging procedures for incorporating interim monitoring devices into clinical research was in the process of codification by regulatory bodies, signifying its acceptance as standard practice in the ethical conduct of clinical research.

1.19 Balancing the Equation

I have provided a brief history of clinical research here, recounting some of the major waves that have lifted and thrust clinical research, and, specifically, interim monitoring procedures forward to their current position. The next task is to demonstrate how these interim monitoring approaches actually work, and, perhaps more important, what they reveal about the population from which the research sample was obtained. Because these concepts are primarily quantitative, a brief review of some basic underlying statistical principles will precede our introductory discussions of the need for monitoring and path analysis.

However, before we begin these new conversations, we should perhaps pause for just a moment to be sure that we have the appropriate ballast to counterbalance our upcoming discussions. The power of mathematics will be demonstrated. However, mathematics, in and of itself, does not provide the answer to the complicated philosophical and ethical considerations that arise in clinical research. Perhaps Norbert Weiner [62] said it best, in the following quote

> One of the chief duties of a mathematician in acting as an advisor to scientists is to discourage them from expecting too much of mathematicians

This remark is prescient for the application of quantitative interim monitoring procedures in clinical research. Making the correct decision to terminate a clinical trial is a multifaceted problem; only some, and not all of those faces are mathematical ones. Speaking of the chi-square test, Bradford Hill said that, "Like fire, it is a good servant, but a bad master." Careful, thoughtful clinical investigators will weigh both the quantifiable and the nonquantifiable alike as they examine each component of the philosophical dilemma that they face. Although the value of a test statistic may trigger a decision-making process, its value in and of itself, is insufficient to end the research endeavor.

In the end, it is our success in balancing the "human equation" that counts. Mathematics is an important consideration, but not the only important consideration in that good effort. This is a perspective that we will strive to maintain in the following chapters.

References

1. Bull JP. (1959) Historical development of clinical therapeutic trials. *Journal of Chronic Disease* **10**:218–248.
2. Malgaigne LF. (1947) *Weuvres Completes d'Ambrosise Paré*, vol. 2. Paris, p. 127.
3. Connolly HM, Crary JL, McGoon MD, et al. (1997) Valvular heart disease associated with fenfluramine-phentermine. *New England Journal of Medicine* **337**:581–588.
4. Clark C. (1997) *Radium Girls: Women and Industrial Health Reform, 1910–1935* Chapel Hill, University of North Carolina Press.
5. Gehan EA, Lemak NA. (1994) *Statistics in Medical Research: Developments in Clinical Trials*. New York, Plenum Medical Book Company.
6. Hill B. (1953) Observation and experiment. *New England Journal of Medicine* **248**:995–1001.
7. Kleinbaum DG, Kupper LL, Morgenstern H. (1982) *Epidemiologic Research: Principles and Quantitative Methods*. New York, Van Nostrand Reinhold Company.
8. Pepall J. (1997) *Methyl Mercury Poisoning: The Minamata Bay Disaster*. Copyright © International Development Research Centre, Ottawa, Canada .
9. Lenz W. (1962) Thalidomide and congenital abnormalities. *Lancet* **1**:45.
10. Moore T. (1995) *Deadly Medicine*. New York, Simon and Schuster.
11. The CAST Investigators. (1989) Preliminary Report: Effect of encainide and flecainide on mortality in a randomized trial of arrhythmia suppression after MI. *New England Journal of Medicine* **212**:406–412.
12. Friedman L, Furberg C, Demets D. (1996) *Fundamentals of Clinical Trials*. 3rd Edition. New York, Springer.

13. Meinert CL (1986) *Clinical Trials Design, Conduct, and Analysis.* New York, Oxford University Press.
14. Piantadosi S. (1997) *Clinical Trials: A Methodologic Perspective.* New York, John Wiley.
15. Yoshioka A. (1998) Use of randomisation in the Medical Research Council's clinical trial of streptomycin in pulmonary tuberculosis in the 1940s. *British Medical Journal* **317**:1220–1223.
16. Medical Research Council Streptomycin in Tuberculosis Trials Committee (1948) *British Medical Journal* **ii**:769–782.
17 Fisher RA. (1925) *Statistical Methods for Research Workers* Edinburg, Oliver and Boyd.
18 Fisher RA. (1933) The arrangement of field experiments. *Journal of the Ministry of Agriculture* 503–513.
19. Moyé LA. (2003) *Multiple Analyzes in Clinical Trials. Fundamentals for Investigators*. New York, Springer. Chapter 1.
20. Amberson JB, McMahon B. Pinner M. (1931) A clinical trial of sanocrysin in pulmonary tuberculosis. *American Review Tuberculosis* **24**:401-35.
21. Diehl HS, Baker AB. Cowan DH.(1938) Cold vaccines: An evaluation based on a controlled study. *Journal of the American Medical Association* **111**:1168-73.
22. Schulz KF. Special features of randomized controlled trials. Located at Website www.icssc.org/PRESENTATIONS (used at workshop/ 23 Special Features RCT.pdf).
23. Hill AB. (1937) *Principles of medical statistics*. London. Lancet, pp. 41-43.
24. Armitage P. (1992) Bradford Hill and the Randomized Controlled Trial. *Pharmaceutical Medicine* **6**: 23-37.
25. Berger VW, Exner DV. (1999) Detecting selection bias in randomized clinical trials. *Controlled Clinical Trials* **20**:319–327.
26. Moseley JB, O'Malley K, Petersen NH, Menke TJ, Brody BA, Kuykendall KH, Hollingsworth JC, Ashton CM, Wray NP (2002) A controlled trial of arthroscopic surgery for osteoarthritis of the knee. *New England Journal of Medicine* **347**:81–8.
27. Pfeffer MA, Sacks FM, Moyé, LA et al. for the Cholesterol and Recurrent Events Clinical Trial Investigators (1995) Cholesterol and Recurrent Events (Cholesterol and recurrent events clinical trial) trial: A secondary prevention trial for normolipidemic patients. *American Journal of Cardiology* **76**:98C–106C.
28. Hill GB. (1983) Controlled clinical trials—the emergence of a paradigm. *Clinical Investigative Medicine* **6**:25-32.
29. Dodge HF, Romig HG. (1929) A method of sampling inspection. *Bell Systems Technical Journal* **8**:613-631.
30. Wald A. (1947) Sequential Analysis. New York: John Wiley and Sons.
31. Armitage P. Sequential Medical Trials 2nd Edition (1975) London. Blackwell.
32. Bross IDJ. (1952) Sequential medical plans. *Biometrics* **8**:188-205.
33. Whitehead JR. (1983) *The Design and Analysis of Sequential Clinical Trials*. Chickester:Ellis Horwood.

34. Ellenberg SS, Fleming TR, DeMets DL. (2003) *Data Monitoring Committees in Clinical Trials: A Practical Perspective*. New York. John Wiley & Sons.
35. Hill AB. (1971) Diseases of treatment. *Public Health.* **85**:107-13.
36. Ford BJ. (1992) Brownian movement in clarkia pollen: a reprise of the first observation. *The Microscope* **40**:235-241.
37. Brown R. (1827) A Brief Account of Microscopical Observations, etc., London
38. Encyclopedia Britannica. http://www.britannica.com/nobel/micro/88_96.html
39. http://www.aip.org/history/einstein/essay-brownian.htm.
40. Jerison D, and Strook D. Norbert Wiener. *Notes of the American Mathematical Society* **42**; 430-438.
41. Weisstein E. Eric Weisstein's World of Physics. http://scieneworld.wolfram.com/physics/BrownianMotion.html.
42. Freudenthal H. (1970) Biography of Norbert Wiener. *Dictionary of Scientific Biography* New York
43. Heybittle JL. (1971) Repeated assessment of results in clinical trials of cancer treatment. *British Journal of Radiology* **44**: 793-797.
44. Pocock SJ. (1977) Group sequential methods in the design and analysis of clinical trials. *Biometrika* **64**: 191-199.
45. O'Brien PC, Fleming TR. (1979) A multiple testing procedure for clinical trials. *Biometrics* **25**:549-556.
46. Lan KKG, DeMets DL. (1983) Discrete sequential boundaries for clinical trials. *Biometrika* **70**:659-663.
47. Lan KKG, DeMets DL. (1989) Changing frequency of interim analysis in sequential monitoring. *Biometrics* **45**:1017-1020.
48. Lan KKG, DeMets DL. (1989) Group sequential procedures. Calendar versus information time. *Statistics in Medicine* **8**: 1191-1198.
49. Whitehead J. (1983) *The Design and Analysis of Sequential Clinical Trials*. New York. Halsted Press.
50. Whitehead J. (1994) Sequential methods based on the boundaries approach for the clinical comparison of survival times. *Statistics in Medicine* **13**:1357-1368.
51. Halperin M. Lan KKG, Ware J, Johnson NJ, DeMets DL. (1982) An aid to data monitoring in long-term clinical trials. *Controlled Clinical Trials* **3**:311-323.
52. Lan KKG, Simon R, Harperin M. (1982) Stochastically curtailed tests in long-term clinical trials. *Communications in Statistics* C, Sequential Analysis **1**:207-219.
53. Lan KKG, Wittes J. (1988) The B-value: A tool for monitoring clinical data. *Biometrics* **44**:579-585.
54. Halperin M, Lan KKG, Ware JH, Johnson NJ, DeMets DL.(1982) An aid to data monitoring in long-term clinical trials. *Controlled Clinical Trials* **3**:311-323.
55. Spiegelhalter DJ, Freedman LS, Balckburn PR. (1986) Monitoring clinical trials. Conditional or predictive power? *Controlled Clinical Trials* **7**: 8-17.
56. Freedman LS, Spiegenhalter DJ. (1989) Comparison of Bayesian with group sequential methods for monitoring clinical trials. *Controlled Clinical Trials.* **10**:357-367.

57. Freeman LS, Spoiegelhbalter DJ, Permar MK. (1994) The what, why, and how of Bayesian clinical trial monitoring. *Statistics in Medicine* **13**:1371-1383.
58. Berry DA. (1993) A case for Bayesianism in clinical trials (with discussion). *Statistics in Medicine* **12**:1377-1404.
59. Carlin BP, Louis TA. (2000) *Bayes and Empirical Bayes Data Analysis* Boca Raton. Chapman and Hall/CRC Press.
60. Parlmar MK, Frittiths GO, Speigelhalter DJ, Souhami RL, Altman DG, van der Scheuren E. CHART Steering Committee (2001) Monitoring of large randomized clinical trials: a new approach with Bayesian methods. *Lancet* **358**:3785-381.
61. US Department of Health and Human Services. (2001) Guidance for clinical trial sponsors: on the establishment and operation of clinical trial data monitoring committees. Draft Guidance FDA.
62. Quoted in D MacHale, *Comic Sections* (Dublin 1993).

2

The Basis of Statistical Reasoning in Medicine

2.1 The Cow in the Investigator's House

Consider the following Russian folk proverb:

> There is a poor farmer who is preparing for the coming winter. Although he had a miserably small and unproductive plot of land and an old cow, his only possession of any value is his house. In reality, it's just a one room shack, but he is nevertheless proud of it and its warm fireplace.
>
> The winter that comes is especially brutal. On one frigid day, the farmer looks out from his warm room and sees his cow, shivering and mooing in the harsh wind outside. He thinks for a moment, and then lets the cow into his house with him. The farmer doesn't approve of what he has done. He doesn't like the idea of letting an animal into his house. He detests the fact that the animal takes up so much of his room. He hates the smell. But he needs the cow to survive. So, in order to live, he gets used to it.*

Health researchers with a nonmathematical background "get used to statistics" in order to survive in research. Appearing to be an amorphous mixture of hard unforgiving mathematics and nebulous concerns about "the freak of chance", statistics can seem to be the worst of everything. A successful businessman relates the following story.

> Sure, I enjoyed mathematics in high school and in college. I actually made the mistake of trying to take a second course in college algebra, and did fine, right up until we got to this thing called "*e* to the *x*". When asked about the ingredients

* Taken from a debate between Soviet Premier Nikita S. Khrushchev and members of the Soviet Politburo in 1962, at the height of the Cuban Missile Crisis.

> that made up this curious entity, the professor said "I can't tell you exactly what "*e*" is, but "*x*" can be anything." That's when I got up, walked right out of the math building, and over to the business school!

To the many healthcare researchers who have no special training in mathematics, statistical thinking is like entering a hall of mirrors. Investigators are interested in proving what they believe, yet statistics seems to focus on disproving what they don't believe. Interpreting multiple endpoints in studies can be particularly complex and troublesome, because although some results are generalizable, others are not. These counterintuitive complications deepen the suspicion that many investigators hold about this mysterious field. Most researchers go into research not because of statistics, but in spite of it.

Clinical investigators need not be experts in statistical computation, but they should be experts in statistical reasoning, that is identifying that relatively small set of circumstances that justify applying results from small samples to large populations. The reliability of the results of a clinical trial rest on the investigators' abilities to separate a true signal of a population effect from the background noise created by the random aggregation of subjects in the sample. A unifying philosophy is critical for interpreting these experiments.

Appropriate statistical reasoning in the presence of interim monitoring is an even more delicate matter. The smaller number of subjects, greater imprecision in the effect size estimates, and the occasional unplanned nature of the analysis combine to complicate, and commonly obfuscate, the best interpretation. Thus, successful implantation of interim monitoring requires not just calculation, but a clear view of that calculation's meaning.

This chapter provides a brief overview of the salient statistical issues in clinical research, serving as a preamble for the discussion of monitoring guidelines that is the main subject of this text. Useful references for more detailed discussions of these issues are available [1,2,3,4].

2.2 Research, Populations, and Samples

The purpose of research is simply to learn, and learning in healthcare requires that we study patients. However, we are faced with the inescapable observation that we cannot study everyone in the population that we wish to understand.

2.2.1 The Tail Wagging the Dog

Consider a researcher who wishes to execute a clinical trial to assess the effect of a new intervention on the overall mortality rate for stroke patients. However, when we press her, we discover that she has a much larger, more audacious goal. There is no question of her clear honest intent to learn about the effect of therapy in these 300 patients. However, she is more focused on applying her results to the U.S. stroke population. Specifically, she wishes to take the findings from her 300 patient study and apply them to the 600,000 patients who have a stroke in the United States each year. If she could have, she would have studied all 600,000 patients, but this was logistically, financially, and ethically* impossible.

Thus, even though she would like to study the entire population, she cannot. What are the implications of this restriction? We know for certain that 600,000 – 300 = 599,700 patients about whom she wants to learn, she in fact will never study. These subjects were never recruited, never randomized, never treated, never followed, and never measured. Yet she claims that she will learn something of value about these unevaluated subjects. This is tantamount to allowing a very small tail to wag a very big dog.

Specifically, studying a small sample drawn from a large population leaves most of the population unstudied and introduces uncertainty. Of course the difficulty is compounded when the researcher ends a study before its scheduled termination time. An early conclusion produces less information available to serve as the basis of generalization to a population. How then can her answer be assured when not just most of the population of interest, but in the case of interim monitoring, much of the sample of interest remains unstudied in her research effort?

The simple and honest response is that there is no guarantee that any sample-based answer is correct. However, there are practices that the investigator can follow that will improve the reliability of the sample-based finding. Specifically, these procedures will make it more likely that the sample's results closely reflects the answer residing in and governing the population from which the sample was drawn. However, even with the use of these modern approaches, the ability to generalize the results from a sample to a population remains limited. Furthermore, the best statistical monitoring effort can not compensate for the weaknesses of a poorly designed and badly executed research effort.

2.3 Three Principles of Sample-Based Research

The *primun movens* of sample-based research is to generalize results from a small sample to a large population. Yet the justification for this extension is not in the motivation, but in the research effort's procedures. Good methodology is greatly assisted by mathematics; however, application of the best research efforts requires an appreciation of sampling that is not so easily quantifiable.

The act of selecting a sample of subjects from the population at large is a combination of science and art. The scientific aspect of sampling is simply the mathematical mechanism used to identify the relatively small number of subjects from the population who will comprise the sample. The art of the process is the ability to tailor the sample to provide clear objective answers to the questions that generated the research. This sculpting ability can be sensitively applied only when the investigator understands what a sample can, and cannot provide.

Because the primary purpose of the research is to learn about a population, the primary purpose of the sample is to *represent* that population. Each individual selected in the sample is selected not just for his or her own attributes, but also to stand in for the many hundreds, thousands, or sometimes millions of patients who

*Many patients, for varied personal reasons, would in all likelihood not have consented to the study.

were never selected for the sample. Therefore, it is simply not enough to measure the germane events that occur in an individual. These occurrences must be evaluated in a way that allows that measurement to represent the unobtainable observations from the unselected members of the population. This difficulty is compounded by the early termination of a study, because the number of subjects on which the termination decision is made is commonly only a fraction of the number of subjects the researchers prospectively declared was the minimum number necessary to complete the effort. In a sense, the "representative value" of a single subject's measurements increases in the early termination environment.

In addition, the investigator recognizes that there is another phenomenon complicating efforts to generalize his results. Different samples, when selected from a single population, will contain different individuals with different life experiences, producing different data. Although the data can be similar across some samples, other samples will reveal marked differences. This sample-to-sample variability is called sampling error. Because the number of subjects on which an early termination decision rests is smaller than the number required to complete the study, the potential effect of sampling error is greater at the interim monitoring point than at the conclusion of the study.

The presence of sampling error raised the question of how likely it is that, despite the investigator's best efforts, the population generated an unrepresentative sample. This possibility is always present in sample-based research, and has an important impact on the use and interpretation of monitoring procedures. Therefore, investigators who wish to draw conclusions from this type of research are obligated to report the degree to which sampling error may have influenced their results.

We may summarize these three principles of good methodological* execution of sample-based research as

Principle 1. Clearly define your question, then select from the population a sample that is representative of the population and whose study will be responsive to your query.

Principle 2. Carry out your sample-based measurements in such a way that the findings in your sample can stand for not just the sample results but can also accurately represent the results that occurred if the study had been carried out in every member of the population.

Principle 3. Accurately measure and report the degree to which sampling error may have misled you.

2.3.1 Analysis Triaging

By their nature, investigators will and must analyze what they believe is illuminating, informative, and interesting. However, they are also obligated to report those findings clearly and in a manner that provides the best interpretation of these re-

* The issue of ethics is central to a productive research effort, but is not the subject of this chapter.

sults. Some analyzes have a clear interpretation and are the most generalizable. Others are not.

Analysis triage can guide the investigators' consideration of generalizable versus hypothesis-generating results. This triage process divides a research effort's evaluations into several clearly defined categories, each of which has a clear interpretation. It is a two-phased process.

The first phase determines if the candidate analysis is confirmatory or exploratory. Is the analysis to be prospectively planned or data driven? The major advantages of prospectively planned analyzes are that the estimates of effect size, confidence intervals, and standard errors are trustworthy. Alternatively, data-driven evaluations, although commonly carried out, and frequently exciting, are less reliable. Post hoc, exploratory results should be executed and reported, but they must be clearly labeled as exploratory. These evaluations require confirmation before they can be integrated into the fund of knowledge of the medical community.

The second level of triage during the design phase of the clinical trial is carried out among the prospectively planned analyzes, dividing them into primary analyzes or secondary analyzes. Primary analyzes are those analyzes on which the conclusions of the trial rest. Each of the primary analyzes will have a prospectively set type I error level attached to it in such a way that the overall (or familywise) type I error rate does not exceed the community accepted level (traditionally 0.05). The trial will be seen as positive, null (no finding of benefit or harm), or negative (harmful result) based on the results of the primary analyzes. It is important to note that a clinical trial can have more than one primary evaluation. If appropriately designed, the study can be judged as positive if any of those primary endpoints is positive.

Secondary endpoints are prospectively declared analyzes in which no attempt is made to control the familywise error rate. Typically, each secondary analysis is typically interpreted at nominal (i.e., judged significant if the p-value is less than 0.05) levels. Secondary analyzes, being prospectively designed, produce trustworthy estimates of effect sizes, precision, and p-values. However, because secondary analyzes do not control the familywise error, the risk of a false positive finding to the population is too great for confirmatory conclusions to be based upon them. Therefore, the role of secondary endpoints is to provide support for the primary endpoint findings, and not to serve as independent confirmatory analyzes.

In the typical clinical trial, there are more exploratory analyzes than there are prospectively declared endpoints, and more secondary endpoints than there are primary endpoints. This is a finding that is consistent with the statement that a small number of key questions should be addressed, accompanied by careful deliberation on the necessity and extent of adjustment for multiple comparisons [5].

We will return to these three principles in Chapters Five through Seven and the impact that they have on monitoring clinical research.

2.4 The Monitoring Complication

The two issues of (1) a sample's ability to represent a population, and (2) the role of sampling error are twin forces that, if not correctly assessed and balanced, can destroy the utility of a clinical study's results. This is, of course, why clinical research must be carefully designed and executed.

Methodological execution requires that the investigator collect a specified number of patients who meet clearly defined criteria and follow them until the end of the study. Only a result of at least a pre-specified magnitude, measured with appropriate precision would be considered positive. However, an investigator who is tasked with monitoring the research effort often finds herself in the position of working to draw a conclusion about a research effort's results before the study is completed. If the study was designed to randomize a precise number of patients and follow them for a pre-specified time period in order to draw an unambiguous conclusion from the study, then doesn't the early termination of a study undermine this well-considered effort?

In many circumstances, the answer is yes. The early termination of a study, if incorrectly managed, will undercut the researchers' efforts to produce a clearly interpretable research result. However, there is a precise set of circumstances in which a clinical research effort may be terminated early and still provide adequate assurances of the result's validity. The clinical investigator's role is to design the research effort so that these circumstances can be created and preserved during the study's execution.

The statistical methodology that governs the monitoring of clinical research must be embedded in the design of a research effort so that its execution, and reliance on its conclusions does not undermine the overall research effort. It must be prospectively detailed, unambiguous, and lead to conclusions that are supported by all of the trial's methodology, in concordance with the three aforementioned principles of sample-based research. These are important constraints, and within these constraints, very few studies can be ended early. Our goal is to understand how the intelligent use of statistical monitoring procedures can aid in the identification of that precise set of circumstances that would lead to the successful and early termination of a clinical research effort.

2.5 The Need for Prospective Design

The need for a clear early statement for the design of a research effort has two general motivations. Although the first is self evident the second is hidden. However, like an iceberg, it is the second, submerged component that is commonly the most damaging when not recognized.

2.5.1 Sample Vision

The first motivation for the prospective design of a study is administrative. Any enterprise, including scientific endeavors, that requires resources needs careful planning to first obtain and then utilize those resources. If one is going to carry out a study evaluating the effect of a genome-drug interaction on short-term lung function (e.g., forced expiratory volume at 1 second (FEV_1)) then the necessary logistics must be in place to produce precise, reproducible measures of FEV_1. If, on the other hand, the purpose of the study is to provide information about the long-term mortality of its participants, then different measurement mechanisms must be in place. These include (1) the legal mechanism to obtain hospital charts and death certificates,* (2) the availability of specialists to determine the cause of death, and 3) the expertise to carry out the appropriate analyzes. Each of these designs is feasible, but requires different resources, and time is required to make these resources available. Clearly, knowing what the study will measure allows the investigator to husband the necessary resources for the study.

2.5.2 Advantages of the Random Sample

A second reason for a prospective design is that early thought must be given to how the sample will be selected. A sample should be selected that, in general, represents the population. However, more specifically, the sample must allow a clear depiction of the relationship that the investigator wishes to illuminate in the population. Ideally, this will involve a random selection mechanism.

We have discussed one use of random mechanisms in clinical research earlier. In Chapter One, discussion focused on attribution of effect, a property that is most directly produced by the random allocation of therapy.† However, the random allocation of therapy is a procedure that is executed after a subject has been selected for sample inclusion. The random selection of subjects is a different mechanism, with a different motivation.

If each population member has the same likelihood of being chosen for the sample, then no member is excluded a priori at the expense of another, and there are no built-in biases against any subject based on that subject's individual characteristics. The procedure that precludes this bias is the *random selection mechanism,* and this process generates a *simple random sample*. The process by which individuals are selected randomly from the population for the sample ensures that every patient in the population has the same opportunity (statisticians say the same constant probability‡) of being selected for the sample.

2.5.3 Limitations of the Random Sample

There are two caveats that we must keep in mind when considering simple random samples. The first is that they are rarely achieved, due to the operation of a set of exclusion criteria. These exclusion criteria are required for logistical and ethical reasons. Sometimes they are used to identify a cohort or collection of individuals that are most likely to demonstrate the relationship that the research is designed to identify.§ Because each exclusion restricts a patient from entering the study based on a characteristic of that patient, the body of exclusion criteria makes the sample less representative of the general population. The inability of most clinical trials to

---

* This aspect of clinical research has become both more important and frustrating as society has become more concerned and restrictive about access to the personal information of individuals.

† Discussed on pages 7–10.

‡ There are more complicated mechanisms that involve random selection. Only the simplest is described here.

§ A fine example is clinical trials that exclude patients who are believed by the investigator to be (1) unlikely to comply with the intervention if it is self-administered over a period of time or (2) unwilling to complete a rigid follow-up attendance schedule.

achieve a sample that even approximates a simple random sample is an important limitation of this research tool.

A second caveat is the incomplete operation of the random chance mechanism in small samples. Because samples exclude most patients from the population, we would expect that a particular sample is not going to represent each of the innumerable descriptive facets of the population. A sample of 1000 patients from a population of 19 million diabetic patients will not provide representative age–ethnic–educational background combinations.* It is asking too much of the sampling mechanism to produce a relatively small sample that is representative of each and every property and trait of the individuals in the population. Thus, the sample must be shaped by the investigator so that it is representative of the population for the traits that are of greatest interest. This contouring process represents a compromise. The sample is created to be representative of some aspects of the population, therefore it is not going to be representative of others. Thus, the resulting sample will have a spectrum of representation, accurately reflecting a relatively small number of traits of the population, and producing inaccurate depictions of others.

Investigators who are unaware of this spectrum, and who therefore report every result from their study as though those results were valid and generalizable simply because they were produced by a random sample, can mislead the medical community. This is a dangerous trap for investigators because it is so easy to collect unrelated but "interesting" data from a study that was itself designed to evaluate a separate question.

For example, consider an investigator interested in determining the change in exercise tolerance in patients with congestive heart failure. After she collects a sample of 300 patients and assesses their exercise tolerance over time, she also queries them about the frequency of hospitalizations for heart failure. In the end, this investigator reports not just the rate of change of exercise tolerance, but also the hospitalization rate of her cohort. She thinks that this is appropriate because she believes that the sample was "representative", and therefore, the hospitalization data are just as reliable as the exercise tolerance data. However, the study was not designed to measure hospitalizations. Patients who were likely to be hospitalized could not meet the entry criteria for exercising, and thus never had the opportunity to enter the study. Hospitalization discharge data was not collected with the same attention to detail as exercise tolerance data. Thus, the ease of collecting data for an evaluation that was not considered during the design of the study, in concert with a sample that was nonrepresentative of hospitalization rates combine to provide a misleading statement about the hospitalization rates for these patients.

This situation is complicated when the study is being monitored for efficacy and safety. During the interim evaluations of this study, the investigator examines both exercise tolerance data and hospitalization rates. As the dataset grows, trends appear and disappear in the dataset for both exercise tolerance and hospitalizations. However, the hospitalization rate interim results can be misleading. The

* For example, if the proportion of subjects who are Hispanic, greater than 65 years of age, and have a graduate school education is less than 1 in 1000, then a sample of 1000 patients is not likely to select any members of the population with these characteristics at all, and the sample will be unresponsive to any questions about this subgroup.

combination of (1) a set of conditions that deselected patients likely to be hospitalized, (2) the inability to obtain good quality data for hospitalization rates, and (3) the random aggregation of subjects in a sample combine to create trends that are not representative and would provide only a misleading perspective on what trends in sample hospitalization rates actually imply for the population at large.

Thus, simple random samples are only representative of the aspect of the population that they were explicitly and overtly designed to measure. Observing a population through a sample is like viewing a complicated and intricately detailed landscape through glasses. It is impossible to grind the glass lens so that every object in the landscape can be viewed with the same sharp detail. If the lens is ground to view near objects, then the important features of the distant objects are distorted. On the other hand, if the lens is ground for the clear depiction of distant objects, then near objects are blurred.

This is a major motivation for concentrating a research effort on a small number of inquiries. By focusing on this short list of questions, investigators are able to choose a sample containing patients with the desired characteristics. However, the wise investigator understands that, by focusing on a small number of prospectively stated questions and selecting a sample that provides representative views of these issues, the sample may not be representative of other characteristics of the population. The sample's results will most likely be generalizable for the questions that it was designed to answer, but not for much else.

2.5.4 Trustworthy Estimators

Measurements on research subjects are combined into estimators that are known by specific names (e.g., means, standard deviations, odds ratios, and relative risks). However, they all have the same function—to provide a reliable estimate of a quantity in the population.

It is easy in this modern computational era to take the reliability of these estimators for granted. However, decades of work were required to identify them and to gain consensus on their use [6].* These estimators have pleasing mathematical properties and are designed to work well in the sampling error environment. It is important to note that they were not designed to remove sampling error. Instead, they channel it into both effect size estimates (e.g., means) and the variability of these estimates (e.g., standard deviations and confidence intervals). If the researcher is also interested in inference (i.e., statistical hypothesis testing), then statistical procedures will channel sampling error into *p*-values and power. Thus, when used correctly, statistical methodology will appropriately recognize and transmit sampling error into familiar quantities that researchers can interpret.

The estimators were designed to perform well in the presence of sampling error. However, for them to function effectively, there can be no other source of

*The idea of repeating and combining observations made on the same quantity appears to have been introduced as a scientific method by Tycho Brae towards the end of the 16th century. He used the arithmetic mean to replace individual observations as a summary measurement. The demonstration that the sample mean was a more precise value than a single measurement did not appear until the end of the 17th century, and was based on the work of the astronomer Flamsten.

random variability. When other, nonsystematic error is present in the research, the estimators on which we rely become unreliable; that is, they no longer measure what they were intended to measure. This most commonly happens in the paradigm of "random research".* Specifically, random research is the circumstance when a sample-based dataset produces an answer to a question that the investigator did not prospectively think to ask.

We already know that one difficulty with the random research paradigm is that the sample was not created to answer the nonprospectively asked question whose answer is suggested by the data analysis. However, a second difficulty lies in the environment in which the estimator is now expected to operate. A sample will, if interrogated often enough, suggest a provocative answer to a question that it was not designed to address. The difficulty with accepting this solution is that other samples from the same population will suggest (1) a different and most times, less provocative answer to this question, and (2) other provocative answers to other questions. These surprise results, being random, appear and disappear across samples; it is not the population transmitting its "signal" through the sample but instead is merely the sampling error "noise" making itself heard.

Estimators function appropriately when they incorporate random data that is gathered in response to a fixed question. They do not perform so well when the selection of the research question is itself random, that is, left to the data. Operating like blind guides, these estimators mislead us about what we would see in the population based on our observations in the sample. The result is a wavering research focus, leaping from one provocative finding to the next, careening wildly about on the random waves of sampling error. Therefore, a primary purpose of the prospective design is to fix the research questions, so that their analysis is well anchored.[7]† This distinction between confirmatory and exploratory analyzes will be particularly important (if not troublesome) in the discussion of interim monitoring for safety in Chapter Seven.

2.6 The Role of Probability and Statistics

The requirement of a sample with its consequent sampling error complicates the interpretation of healthcare research. The implications of sampling error are profound, forcing the investigator to predict or estimate what would happen in the population based on what is observed in the sample. Sampling error can distort the view of the population, and if the research questions are not insulated from the effects of sampling error, this source of variability can wreck the ability of the estimators to provide any useful information at all.

Because the researchers' efforts to predict and estimate population effects from sample findings (1) are quantitative, and (2) must acknowledge and incorporate the notion of random error, it is only natural that they incorporate statistics. Statistics focuses on the ability to first estimate a population quantity based on data that is obtained from a sample, and then, if necessary, infer a population relation-

---

* This issue is discussed in Chapter 2 of [4].

† The problems with midstream changes in analysis plans occasionally rises to the level of public awareness. This most recent was a statement by an FDA scientist.

ship (e.g., treatment-induced lower cumulative fatal and nonfatal stroke rate) from observations in the population. We will now turn our attention to the basic idea behind statistical inference.

2.7 Statistical Inference

The preceding discussion of the use of the sample has prepared us for the notion of statistical inference. The purpose of drawing the sample is to learn something of value about the population from which the sample was obtained, that is, to infer from the sample to the population. Statistical inference is the process by which that inference is carried out.

We already understand some of the pivotal steps in the inference process. The researcher must choose the appropriate estimators. The environment in which these estimators operate must not be shaken by the perturbations generated by changes in a research protocol that are induced by findings in the data. These concepts are central to the ability of these estimators to estimate what they were designed to measure.

Statistical inference focuses on what to do with these estimators once they have been obtained. Although the concept of statistical inference in healthcare is well accepted, there have been important and continuous disagreements as to how this inference should be carried out. The use of formal hypothesis testing, a tradition that has strong roots in the medical research community, took root in the 1940s and remains a central component of healthcare research. This paradigm involves the construction of null and alternative hypotheses, and ultimately the generation of a *p*-value. The groundswell of enthusiasm for this perspective has been tracked and discussed [8]. In fact, the hypothesis testing scenario has become so popular that the notion of statistical inference and statistical hypothesis testing have become synonymous in healthcare research. However, there are other approaches to drawing conclusions from a sample to a population that do not involve formal hypothesis testing that have demonstrated themselves to be worthy competitors.

2.7.1 Confidence Intervals

Although there has been a 60 year tradition of carrying out formal statistical hypothesis testing, an influential community of epidemiologists has developed a continuous and formidable resistance to its application to healthcare research. The appearance of misleading research results from studies that have abused the hypothesis testing scenario, in concert with the combination of sample size, effect size, and variability into one number has caused many in epidemiology to eschew the *p*-value for the confidence interval[9].

The confidence interval provides important and useful information about the role that sampling error plays in the generation of the result. Incorporating the point estimate and its standard error, the confidence interval provides a readily interpretable assessment of the point estimate's precision.

This concept can be illustrated by an example from a recent clinical study. The Heart Outcomes Prevention Evaluation (HOPE) trial was designed to assess the effect of the ACE-i therapy ramipril on clinical measures of cardiovascular disease [10]. It was well-designed, and executed in accordance with its protocol (i.e., the research effort was *concordantly executed*). At its conclusion, one of its major findings was the effect of ramipril on the combined measure of myocardial infarction (MI), stroke, or cardiovascular (CV) death. The relative risk for this effect was 0.78* and the 95% confidence interval was 0.70 to 0.86[11].

This 95% confidence interval draws attention to the range of possible values of the relative risk in the population from which the HOPE sample was drawn. A common interpretation of the confidence interval is conveyed by saying that it is likely that the value of the true relative risk lies somewhere in this 0.70 to 0.86 range.†

Although useful, sole reliance on the confidence interval has its opponents. One criticism of the confidence interval approach is that it does not easily lead to dichotomous decisions (e.g., is the therapy effective). Some workers rely on whether the confidence interval contains the value 1 (signifying no therapy effect) as evidence that the therapy is unlikely to be effective in the population; therefore this is tantamount to carrying out a hypothesis test, a procedure that the worker may be attempting to avoid. Finally, the confidence interval, much like the estimate of the relative risk itself, is only accurate insofar as the research environment is a concordant one. Data-based changes in the protocol undermine the confidence interval estimate as easily as they destabilize the estimate of the relative risk.

2.7.2 Bayesian Procedures

An alternative to the traditional hypothesis testing paradigm is the implementation of Bayes procedures. Their underlying philosophy is distinct enough from the standard (or frequentist approach) to statistical inference that many now view the two perspectives as polar opposites, and the literature is replete with vigorous debates between the zealots of each philosophy. However, here we will steer well clear of these controversies, contenting ourselves with a brief review of each approach.

2.7.2.1 Classical Statistics (the "Frequentists")

Classical statistics is the collection of statistical techniques and devices that evaluate the accuracy of a technique in terms of its long-term, repetitive accuracy [12]. The conclusion from any particular research effort may be wrong. However, if the experiment were repeated many times, the application of classic hypothesis testing procedures would produce the correct answer most times. This concept of the over-

* The relative risk of the effect of therapy demonstrated a 1 – 0.78 = 0.22 or 22% reduction in the incidence of the combined endpoint associated with the use of ramipril.

†This is not the most accurate definition, because it suggests that the variability is associated with the population relative risk which, in this paradigm, is constant. The sample-to-sample variability is associated with the location of the 95% confidence interval and whether it contains the population relative risk. The most accurate interpretation of the HOPE-generated confidence interval is as follows. If there were 100 samples obtained (in this case, this would mean that 100 HOPE studies were performed), each with its own confidence interval, then 95% of these confidence intervals would contain the true population relative risk. Of course, with only one study, and one confidence interval, we do not know one way or the other whether this confidence interval contains the true value of the relative risk.

all accuracy of the procedure buoys the confidence of the classical statistician, even though the wrong conclusion may be obtained from any particular experiment. However, for researchers who have much of their time (and sometimes, the taxpayers or stockholders' money) bound up in an important research effort, this observation can produce a small shock. The realization that sampling error can lead to an erroneous result in even an expensive, well-designed, and well-executed research effort is not very comforting.

We have already seen this principle in operation. The use of the confidence interval to estimate a relative risk does not guarantee that any particular experiment will generate a confidence interval that contains the true value of the relative risk. Instead, the underlying principle provides an assurance that, in the overwhelming number of samples (95% of these samples for a 95% confidence interval), the confidence interval will contain the population relative risk.

Another frequentist characteristic is the focus on not just what has occurred, but what has not occurred in a research program. How these non-occurrences are handled can have a dramatic impact on the answer to a scientific question, and preoccupation with them can bedevil the investigator .

2.7.2.2 The Bayesians

Like classical statistics, Bayes theory is applicable to problems of parameter estimation and hypothesis testing. The Bayesian formulation is based on a principle, termed the likelihood principle, which states that a decision should have its foundation in what has occurred, not in what has not occurred.

Like frequentists, Bayesians are interested in parameter estimation and hypothesis testing. Bayesians estimate the population parameter θ of a distribution (just as frequentists do). However, unlike frequentists who believe that the parameter is constant, Bayesians treat the parameter as though it itself has a probability distribution. This is called the *prior distribution,* signified as $\pi(\theta)$.

Once the prior distribution is identified, the Bayesian works forward, next identifying the probability distribution of the data given the value of the parameter. This distribution is described as the *conditional distribution* (because it is the distribution of the data conditional on the value of the unknown parameter) and is denoted as $f(x_1, x_2, x_3, ..., x_n \mid \theta)$. This step is not unlike that of the frequentist. When attempting to identify the mean change in blood pressure for a collection of individuals, both the frequentist and the Bayesian may assume that the distribution of blood pressures for this sample of individuals follows a normal distribution with an unknown mean, whose estimation is the goal. However, the frequentist treats this unknown mean as a fixed parameter. The Bayesian assumes that the parameter is not constant, but changes over time. Its distribution is called the *prior distribution.*

The Bayes process continues by combining the prior distribution with this conditional distribution to create a *posterior distribution*, or the distribution of the parameter θ given the observed sample, denoted as $\pi(\theta \mid x_1, x_2, x_3, ..., x_n)$. From the Bayes perspective, the prior distribution reflects knowledge about the location and behavior of θ before the experiment is carried out. After the experiment is executed, the researcher has new information in the form of the conditional distribution.

These two sources of information are combined to obtain a new estimate of θ. To help in interpreting the posterior distribution, some Bayesians will construct a loss function, which identifies the penalty that they pay for underestimating or overestimating the population parameter. Bayesian hypothesis testing on the value of the parameter is based on the posterior distribution.

The Bayesian approach to statistical analysis makes unique contributions. It explicitly considers available prior distribution information, and allows construction of a loss function that directly and clearly states the loss (or gain) for each decision. However, the requirement of a specification of the prior distribution can be a burden if there is not much good information about the parameter to be estimated. Similarly, the choice of the loss function can be difficult to justify from a clinical perspective.

Several interim monitoring procedures have been developed that reflect the Bayes perspective about which we will have more to say in Chapter Eight.

2.7.3 Hypothesis Testing Paradigm

The scientific method, easily recognized as the driving force motivating research efforts, begins with an idea. Investigator conceived and formulated, this idea is commonly an affirmative one; for example, a new class of drug is an acceptable alternative to coumadin for the prevention of stroke in patients with atrial fibrillation (AF). This clinical postulate either represents scientific truth or it does not. In order to determine the accuracy and applicability of the researcher's concept, the investigators carry out an experiment. During the design of this research, the clinical hypothesis is converted into one (or a collection of) statistical hypotheses.

The scientific method begins with a hypothesis or initial idea that the researcher hopes to prove. If θ_A is the cumulative stroke rate in the active group, and θ_C is the cumulative stroke rate in the control group, then the investigator believes and states his clinical hypothesis as $\theta_A < \theta_C$. However, statistical hypothesis testing commonly begins with a hypothesis that the researcher hopes to disprove or nullify. Thus, the investigator who believes the new class of drugs is beneficial will commonly state a *null hypothesis*. In the current example, the null hypothesis is that patients who are assigned to the new class of drugs will have the same stroke rate as those patients who were assigned to coumadin. Thus, the statistical null hypothesis is not that $\theta_A < \theta_C$ but that $\theta_A = \theta_C$. It is this null hypothesis that the investigator wishes to disprove or nullify with the experiment's results.

The reason for this change of emphasis from a positive clinical hypothesis to a null statistical hypothesis deserves some discussion. The investigator cannot be blamed for his first impression that, by being forced to turn away from proving an affirmative hypothesis to disproving a null one, he has lost the intellectual initiative. However, the investigator must recognize that he himself has chosen to be involved in an act of nullification. Specifically, he has chosen to nullify the current approach that is used to prevent post-AF thrombotic events.

Prior to the investigator's research, the current, accepted standard of care in the medical community is that coumadin is the best outpatient therapy available to reduce the stroke rate in patients with AF. By believing that the new class of drugs is better than coumadin, clearly an affirmative concept, the investigator announces his nonacceptance of the assertion that the accepted standard of care is optimum. He wishes to nullify this belief, and he will do that by designing a trial that demonstrates the effectiveness of the new class of drug.

The design of the clinical trial is central to this process. Patients are randomized to active or control group therapy in order to minimize differences between the groups. Investigators endeavor to ensure that patients are treated similarly across the two groups. They work to reduce differences in compliance with medication between the groups. The investigators determine the occurrence of clinical endpoints without knowledge of the patient's assigned therapy. This system is constructed so that, if the current standard of care is correct, there will be no differences in the cumulative stroke rates between the two groups. Thus, if there are important differences in the stroke incidence rates, they can be due to only two reasons: (1) the freak of chance produced by the random aggregation of patients in the research sample, or (2) the therapy actually made a difference.

Therefore, the null hypothesis is merely a mathematical characterization of the current practice of medicine. It is consistent with an underlying theme of the clinical trial; that is, if the therapies being tested have the same effect on the clinical endpoint, then the data and the trial support the state of the art, or the null hypothesis.

When the findings are more extreme, they are commonly described as "unlikely to be due to chance alone". This means that the sample-to-sample variability is too small to serve as the only explanation for the large difference in the stroke rates between the two groups. This statement has come to be encapsulated in the *p*-value.

2.7.4 P-Values

The *p*-value is a measure of sampling error. It is simple in concept, but its long and complex history is undeniable. Before we put their use in context, let's first discuss what *p*-values are supposed to be, and then acknowledge what they have become.

When the positive conclusions of a well-designed, concordantly executed research program are placed in her hand, the researcher must acknowledge two possible explanations for the results. The first is that the sample truly represents the findings of the larger population. However, the second explanation is that the sample's results are due to sampling error and do not represent the population.

This second explanation is motivated by what sampling error can produce in the absence of a real treatment effect in the population. In this situation, the positive research finding does not accurately reflect relationships in the population. Instead, the positive sample findings are unique to the sample. They are not seen in, nor are they representative of, the population. Just as a "population" of 1000 fair coins when flipped could produce a "sample" of five coins that all showed "heads", a population that has no effect can produce a sample which contains, just through the play of chance, an important "treatment effect".

Because sampling error is always present in sample-based research, we can never know for sure whether a sample's results are representative of the population or just due to the play of chance. The *p*-value simply measures the likelihood that sampling error has produced a positive result in the research sample of the investigator. If sampling error produced the research result, the researcher would be vestigator. If sampling error produced the research result, the researcher would be wrong in concluding that the effect seen in her sample represents a true finding in the population. The *p*-value is the probability that a population in which there is no effect would produce a sample that demonstrates an effect.* The smaller the *p*-value, the less satisfactory is the explanation that sampling error explained the results, and the more likely a truly representative research result was identified.

The idea of the *p*-value and significance testing is based on the work of the agricultural statistician Ronald Fisher.† As he worked through the design and analyzes of agrarian experiments in the 1920s, he stated that, if there was a greater than five percent chance that a population that had no positive findings produced a sample with positive findings, the positive findings in the sample should be discarded because the likelihood that they were due to the random, meaningless aggregation of events was too great [13,14].

This was the beginning of "significance testing", and the "$p < 0.05$" concept. There is no deep mathematical theory that points to 0.05 as the optimum type I error level—only tradition. The rise and pre-eminence of the 0.05 level has its roots less in science and more in the "sociology of science", as 1940s journal editors and senior grant reviewers struggled with differentiating worthy scientific results from second-tier ones [15,16].

Unfortunately, many researchers have substituted the 0.05 criterion for their own thoughtful, critical review of a research effort, and this replacement has led to uninformed research interpretation. Poole [17] pointed out that the mechanical reflexive acceptance of *p*-values at the 0.05 level is the nonscientific, easy way out of critical and necessary scientific discussions. For example, highly statistically significant effects (i.e., results associated with small *p*-values) have been produced by small, inconsequential effect sizes. In other research efforts, small *p*-values themselves were rendered meaningless when the assumptions on which they had been computed were violated. In addition, there is the observation that statistical significance may not indicate true scientific, biological, clinical, or economic significance [18,19, 20, 21, 22].

The reduction of a complex research endeavor's result to a single *p*-value is perhaps at the root of the inappropriate role of significance testing. This condensation effort may be due to the fact that the *p*-value is itself constructed from several constituents. Sample size, effect size, and effect size variability are important components of the *p*-value and are directly incorporated into the *p*-value's formulation. However, in reality, what is produced is not a balanced measure of these important contributory components, but only a measure of the role of sampling error as a possible explanation for the results observed in the research sample. Thus, *p*-values are deficient reflections of the results of a research effort, and must be supplemented with additional information (the research effort's concordance,‡ sample

---

* The α error rate is the type I error rate that is set before the research begins. The *p*-value is the measure of α that is based on the result of the research.

† This is the same Ronald Fisher whose contribution of the tool of randomization was discussed in Chapter One.

‡ Concordance is the desirable property of research that derives from the tight match between the research execution and the plans for its execution as stated in the research protocol.

size, effect size, and effect size precision) in order for the study to receive a fair and balanced interpretation [23]. The investigator must jointly consider these measures when interpreting a research endeavor's results.

2.7.5 Statistical Power

Consider a research effort that is designed to identify the effect of a therapeutic intervention in a sample of patients. As we saw previously, sampling error can produce from a population in which there is no therapeutic effect of interest a sample in which there is a significant treatment effect. This misleading sample is generated by the random and unpredictable selection of subjects in the sample, that is, by chance alone.

However, the influence of sampling error can be equally insidious when the study results are not positive. After all, it is quite possible that a population in which there is a treatment effect of interest may produce a sample in which there is no effect. In this case, the researcher is compelled to conclude that, because his research sample produced no effect of interest, the therapy is not effective for its studied use. However, this would be a false result because this sample that produced the null finding was produced by chance alone. This is called a *type II error*. The probability that a population in which there is an important treatment effect also produces a sample containing that same effect is the power of the study, and may be computed as simply $Power = 1 - \boldsymbol{P}[Type\ II\ error]$.

Unlike the case of the *p*-value, where there has been a strong tradition of setting the threshold at 0.05, the minimal acceptable power for a study has been a standard that has changed over time. Acceptable power levels can extend up to and sometimes exceed 95%. Rarely, however, is a study acceptable that is based on a power level of less than 80%.

Which one of the type I error or type II error the reader of a well-designed and concordantly executed* study should track depends on the findings of the research. If the study results are positive, then the reader must focus on the likelihood that a false positive study could be produced through sampling error. This is a concern that focuses on the *p*-value. If, on the other hand, the study findings are null, the reader turns her attention to the power of the study. If the power of the study is high, she may assume that it is unlikely that a population in which the research effect was important would produce by chance alone a sample in which the research effect was absent.

During the design phase of the study, the researcher will not know which of these two errors may occur, so he must design the study with a priori concern for each of these errors.

---

* Of course, if the research is not designed well, or is executed poorly, then these measures of sampling error can be corrupted and therefore, inaccurate. One learns how well the research was designed and executed from an examination of the methodology section of the manuscript. See Chapter 4, Section 3.2 Systematic Reviews from Moyé [22].

2.8 Conclusions

Study results from undisciplined research efforts can produce provocative findings with no lasting value. The same tendency, left unchecked, will also plague the results produced from the interim monitoring on clinical research. These intermediate results, based on smaller samples, powered by small *p*-values, can with loud voice point the medical community in the wrong direction.

As long as research is based on drawing a sample from a population, sampling error will play a role in the product of that research. Two important precepts to follow are (1) clearly define your question, (2) select from the population a sample that is representative of the population, and (3) develop a clear a priori protocol and follow that protocol during the conduct of the research. Adhering to these principles will contain the extent and the limit the role of sampling error as an explanation of the research results, be they interim results or the final results of the study.

References

1. Friedman L, Furberg C, Demets D. (1996) *Fundamentals of Clinical Trials.* 3rd Edition. New York, Springer.
2. Meinert CL. (1986) *Clinical Trials Design, Conduct, and Analysis*, New York, Oxford University Press.
3. Piantadosi S. (1997) *Clinical Trials: A Methodologic Perspective.* New York, John Wiley.
4. Moyé L. (2003). *Multiple Analyzes in Clinical Trials: Fundamentals for Investigators.* New York. Springer.
5. Proschan MA, Waclawiw MA. (2000) Practical guidelines for multiplicity adjustment in clinical trials. *Controlled Clinical Trials* **21**:527–539.
6. Plackett RL. (1958) The principle of the arithmetic mean. *Biometrika* **45** 130–135.
7. Harris G. (2004) Merck says it will post the results of all drug trials. *New York Times.* September 6, 2004. C4.
8. Gigerenzer G, Swijtink Z, Porter T, Dasxton L, Beatty J, Kruger L. (1989) *The Empire of Chance* Cambridge. Cambridge University Press.
9. Moyé L. (2000). *Statistical Reasoning in Medicine. The Intuitive P-value Primer.* New York, Springer.
10. The Heart Outcomes Prevention Evaluation (HOPE) Study Investigators. (1996) The HOPE (Heart Outcomes Prevention Evaluation) Study: The design of a large, simple randomized trial of an angiotensin–converting enzyme inhibitor (ramipril) and vitamin E in patients at high risk of cardiovascular events. *Canadian Journal of Cardiology* **12**:127–137.
11. The Heart Outcomes Prevention Evaluation Study Investigators. (2000) Effects of angiotensin–converting enzyme inhibitor, ramipril, on cardiovascular events in high–risk patients. *New England Journal of Medicine* **342**:145–53.
12. Berger JO. (1980). *Statistical DecisionTheory. Foundations, Concepts and Methods.* New York, Springer-Verlag.
13. Fisher RA. (1926) *Statistical Methods for Research Workers.* Edinburg. Oliver and Boyd.

14. Fisher RA. (1933) The arrangement of field experiments. *Journal of the Ministry of Agriculture*. 503 - 513.
15. Goodman SN. (1999). Towards evidence–based medical statistics. 1: The *p*-value fallacy. *Annals of Internal Medicine* **130**:995–1004.
16. Gigerenzer G, Swijtink Z, Porter T, Dasxton L, Beatty J, Kruger L. (1989) *The Empire of Chance*. Cambridge, Cambridge University Press.
17. Poole C. (1987) Beyond the confidence interval. *American Journal of Public Health* **77**:195–199.
18. Walker AM. (1986) Significance tests [sic] represent consensus and standard practice (Letter). *American Journal of Public Health* 76:1033. (See also journal erratum **76**:1087.
19. Fleiss JL. (1986) Significance tests have a role in epidemiologic research; reactions to AM Walker. (different views) *American Journal of Public Health*.**76**:559–560.
20. Fleiss JL. (1986) Dr. Fleiss response (Letter) *American Journal of Public Health* **76**:1033–1034.
21. Walker AM. (1986) Reporting the results of epidemiologic studies. *American Journal of Public Health***76**:556–558.
22. Thompson WD. (1987) Statistical criteria in the interpretation of epidemiologic data (different views) *American Journal of Public Health* **77**: 191–194.
23. Moyé L. (2004) *Finding Your Way in Science: HowYou Can Combine Character, Compassion, and Productivity in Your Research Career*. Vancouver. Trafford Press.

3

Probability Tools for Monitoring Rules

Commonly, the application of statistical procedures to a clinical research endeavor ends with a hypothesis test. Clinical monitoring doesn't end with a test statistic, but begins with one. The investigator holds the value of a test statistic at a particular point in time in her hand, and then, using this test statistic, attempts a prediction of the probability of a future outcome. Because this future event is only one of many possible events, probability is central to its evaluation.

Our purpose here is to gain the intuition in probability that we need in order to be able to assess the interim results of a clinical research activity. That requires that we review the concept of actions that produce unpredictable events, and acquire a basic understanding of how probability tools operate on these events. Because the reader is most likely uninterested in the inner workings of mathematical models, the computations involving those models are not reproduced here.* Therefore, although probability has a deep history and is enveloped in a lush literature, we will not overstay our welcome.

3.1 Using Probability to Monitor Research

Probability computes the likelihood of events that occur from actions. An *action* is a maneuver that results in an uncertain outcome (e.g., flipping a coin). These outcomes or events are identifiable, but the knowledge of exactly which outcome will be produced by the next action is not available a priori.

Probability is an important tool for the clinical monitor who, with an interim result available, wishes to learn if the study will ultimately be positive. At this interim point in time, there are many possible future results. Commonly the types of study results are collected into classes or categories of results. One example is the collection of results that are all positive, that is, that show a beneficial effect of the tested clinical intervention.

Even though the clinical investigator will commonly not be the scientist who carries out the actual probability computation, he bears the key responsibility of constructing and articulating the clinical event whose probability holds his inter-

* In all likelihood, the reader will have a statistician or some other quantitative specialist that will carry out the necessary computations.

est. This is not a trivial matter, and often requires a good deal of thought. Although the investigator can be excused from knowing the details of the probability models, he is fully accountable for understanding the event whose probability he has requested to be computed.

3.2 Subjective Probability

Probability is ubiquitous and is a concept that is frequently incorporated in conversation. Commonly, it is not used mathematically, but is instead used merely as a vehicle to convey an opinion. For example, an attending physician who is discussing a case with medical students may conclude rounds with the comment, "We need to wait for all of the lab analyzes to return, but the patient, in my view, probably does not have subacute bacterial endocarditis." Rushing away to a staff meeting, he stops to receive a brief telephone call from his daughter that leads him to believe that, "She probably is not going to find a job this summer." During the meeting, he learns that the faculty parking lot, "Will probably be replaced," by a new campus building. At the conclusion of the meeting, he receives another call, this time from the car mechanic who says his wife's car, "Will probably not be repaired," in time for the weekend.

Each of these four examples demonstrates the use of a type of amorphous nonquantitative probability. No formal evaluation was embedded in the preceding four probability assessments, yet each statement conferred an assessment of the most likely outcome of an action. These probabilities are referred to as *subjective probabilities.* They are based on one's personal beliefs, cultural opinions, and attitudes.

Subjective probabilities can be complicated, representing a diffuse mixture of qualitative as well as quantitative components. In the aforementioned examples, there was no satisfying, intuitive mathematical device that would produce an assessment for the attending physician of the likelihood of the events of interest. A natural and human response on his part was to fill this gap with his own experience.

It is important not to disparage the notion of subjective probability. To a great extent, subjective probability is used in monitoring a clinical research effort. As the monitor, we may receive a good deal of quantitative information. Some of it is rigid and formal, for example, whether a test statistic has crossed a threshold or boundary value. Other information is based on qualitative evaluations for example, the findings of other endpoints in the research effort, or the results of other investigations that are conducted in parallel with the one in question. Ethical concerns are certainly nonquantitative and play a central role in the review process.

It is up to us as physicians and scientists monitoring the study to integrate this information from multiple sources correctly. We do this by first combining the mathematical and non-mathematical information. This mixture will leave gaps in our information; these gaps demonstrates the complexity of the decision process. We react to these missing components by filling them in with pieces of ourselves, using not just our intuition, knowledge, training, experience, and expertise but also our personality, biases, hopes, and fears. This is the heart of subjective probability. Although this is all that we will say about it, its role in assessment of interim results is undeniable.

3.3 Probability as Relative Frequency

As useful and as necessary as subjective probability is, a main focus of the monitoring effort for clinical research is on the mathematical assessment of probability. This is most directly related to the action–outcome scenario. We define an *action* as a sequence of occurrences resulting in an unknown outcome. A simple action is the flip of a coin. A more complex action would be the combination of eventualities that produces a particular interim result in a clinical study.

Outcomes or *events* are the results of an action. The possible outcomes of this action are identifiable, but precisely which outcome will occur is unknowable. The collection of all possible events is the universe or *sample space* of events. Of the sample space of events, we are concerned about the occurrence of events of interest. We may define probability as

$$\boldsymbol{P}[\textit{event of interest}] = \frac{\textit{frequency of all events of interest}}{\textit{frequency of all possible events}}.$$

For example, we are interested in flipping a coin twice. Let H represent a head, and T represent a tail. We are interested in the event where the two flips provide different results. Our universe of events is $\{H, H\}$, $\{H, T\}$, $\{T, H\}$, $\{T, T\}$. The event of interest is $\{H, T\}$, $\{T, H\}$. We compute the probability as

$$\boldsymbol{P}[\textit{alternating results}] = \frac{\textit{frequency}\left(\{H,T\},\{T,H\}\right)}{\textit{frequency}\left(\{H,H\},\{H,T\},\{T,T\},\{T,H\}\right)}.$$

In this case the frequency is just the number of events. Thus,

$$\boldsymbol{P}[\textit{alternating results}] = \frac{2}{4} = \frac{1}{2}.$$

The two critical aspects of computing probability are (1) identification of the event of interest and (2) the ability to count or "measure" them*. These can both be complicated, so we will spend time on each of them.

Probability is a mathematical function, but is unlike any mathematical function that you may have been exposed to in college algebra or calculus. Typically, we think of mathematical functions that are simple mappings, that is, they map or convert one number to another. A simple example of this is the function $f(x) = 5x$. This function merely defines a rule that governs the conversion of one number to another. In this case, the rule is to take the first number or argument, and

* Assigning an appropriate measure to uncountably many paths of Brownian motion was the central contribution of Norbert Weiner to Brownian motion predictions that has become so useful in applying Brownian motion to clinical research monitoring.

multiply it by five. A more complicated function is $f(x,y) = x/y^2$. Here, the function maps two arguments, x and y, to a single number $f(x,y)$. In this case, if x is a patient's weight in kilograms, and y is his height in meters, then $f(x,y)$ represents his body mass index or BMI.

Probability is a function with a very different argument. The argument of probability is not always a real number, but is best thought of as a set, or a collection of items.

3.4 Decomposing Events

We will have many events for which probabilities are required in modern monitoring procedures in clinical trials. The problem that faces the investigator is the probability computation of a complicated event or outcome (Figure 3.1).

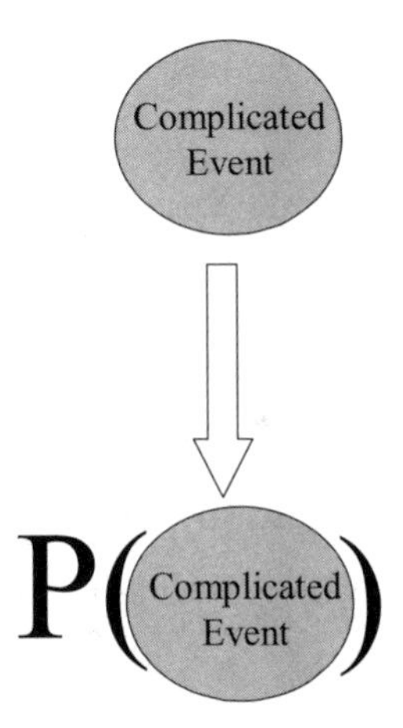

Figure 3.1. The desire to compute the probability of complex monitoring events. **P( )** signifies the probability function.

This complicated event may be very easily stated, for example, the event that the study being monitored will produce a negative or harmful result.* This is an easy event to conceptualize. However, exactly how one goes about computing the probability of such an event is unclear, because the random aggregation of clinical circumstances that produce this event seems both innumerable and unpredictable.

One important way that a probabilist approaches complex problems is through the strategy of deconstruction and construction. Essentially, a complicated problem is disassembled into simpler components. These simple components, being

* In this text, a positive study is a study where the intervention produces a beneficial effect. A negative study demonstrates a harmful effect, and a null study produces neither benefit nor harm. Every attempt will be made to ensure that the research context is clear in order to reduce any confusion through the use of these terms.

easier to understand, are studied and understood. This understanding leads to a knowledgeable reconstruction of the original complicated problem and its solution.

In the circumstance of monitoring clinical research, this approach translates into taking apart the complicated event whose probability is sought into simpler events. This deconstruction is both straightforward and reproducible. The simple events have probabilities that are more easily computed. Then, using the rules of probability, the solution to the original problem is assembled from the probability of the individual components. The sequence is (1) event deconstruction, (2) probability computation, and (3) probability reconstruction. (Figure 3.2).

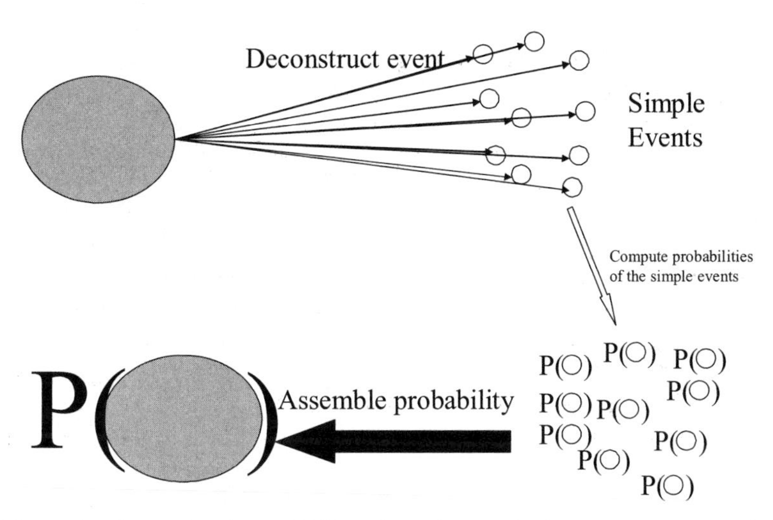

Figure 3.2. Computing the probability of complex monitoring events. The complicated event is deconstructed to a collection of simple events whose probabilities are easily computed. These probabilities are then reassembled into the probability of the original, complex event.

Thus, in order to carry out these operations, we need to know how to deconstruct a complex event into a combination of simpler events whose probabilities we can easily find.

3.4.1 Combinations of Events

Consider a collection of events $A_1, A_2, A_3, \ldots, A_n$ as events whose probabilities can be computed. For example, consider a hospital that has 100 patients from which we would like to select a small sample of five patients to determine their insurance status. Let A_1 be the event that the first patient is part of the sample of size 5. A_2 is the event that the second patient is in the sample, and A_3 is the event that the third patient is in the sample. Finally, A_{100} is the event that the 100th patient is in the sample.

In this scenario, it makes sense for us to be able to compute the probabilities of each of these events (which we will do shortly). However, there are many other events that could occur that are of interest as well. For example, the event that both patient 1 and patient 2 are in the sample of size 5 is a possibility whose probability we might like to compute. Another event of interest is that either patient 1 or patient 2 is in the sample of size 5. Also, the event that neither patient 1 nor patient 2 is in the sample of size 5 would be interesting. These latter events, although more complicated than the events signified by $A_1, A_2, A_3, \ldots, A_{100}$, are nevertheless related to them. We will describe these relationships, and then show how to compute the probabilities of these events.

3.4.2 Joint Occurrences

Monitoring rules in clinical trials commonly focus on the computation of events that involve the joint occurrences of events (e.g., computing the probability that a measure of effectiveness exceeds a value when 50% and 75% of the trial's total follow-up time has elapsed). Therefore the ability to successfully understand and manipulate probabilities of joint occurrences is critical to understanding these monitoring procedures. The event that A_1 and A_2 occur means that both events must be satisfied. Using the insurance example, a joint event would be that both patient 1 and patient 2 must be in the sample of size 5. The inclusion of either is not sufficient; both must be present. We signify this compound event as $A_1 \cap A_2$, and read it as the event A_1 and A_2. The compound event that requires both events must occur is known as an *intersection.*[*] There are many possible intersections from our collection of events $A_1, A_2, A_3, \ldots, A_{100}$. Examples are $A_{19} \cap A_{23}$ (the 19th and 23rd patient are in the sample of size 5), and $A_{94} \cap A_{95} \cap A_{96}$ (the 94th, 95th, and 96th patients are in the sample of size 5). These are called the probabilities of joint occurrences, or *joint probabilities*.

How we compute probabilities of these joint events depends on the relationships between the events. Events can be *mutually exclusive.* If they are not mutually exclusive, they can be *independent* or *dependent.* The definitions of these properties are straightforward.

3.4.3 Mutual Exclusivity

Two events E_1 and E_2 are mutually exclusive if the occurrence of one makes the occurrence of the other impossible. For example, if E_1 is the occurrence of a head on the single flip of a coin, and E_2 is the occurrence of a tail then E_1 and E_2 are mutually exclusive. This is because a single flip of a coin produces one and only one result.

A moment's reflection reveals that intersections of mutually exclusive events are impossible. In the coin-flipping scenario where E_1 and E_2 are mutually exclusive, we may write $\boldsymbol{P}[E_1 \cap E_2] = 0$. Thus, computing the probability of joint events is trivial when the events are mutually exclusive.

Returning to the hospital scenario, the answer to the question, "Are A_1 and A_2 mutually exclusive?" is no, because both patient 1 and patient 2 can each be part

[*] Probabilists frequently state this as "A_1 intersect A_2".

of the sample of five patients. The next question that we ask is are A_1 and A_2 independent.

3.4.4 Independence

When events are not mutually exclusive, we must examine the events carefully for other useful properties. Among the most useful is the notion of independence. Events can be independent or dependent. Events are independent if the occurrence of one event tells us nothing about the occurrence of the other. Dependent events are events where the occurrence of one event allows us to adjust the probability of the occurrence of the other event.

Most of us have a general understanding of the concepts of dependence versus independence. However, although these two properties are somewhat intuitive, we need to elaborate on their fundamental characteristics. The descriptors "independence" or "dependence" are properties of a relationship. We don't ask if the occurrence of macular degeneration is "independent." We do ask if the occurrence of macular degeneration is dependent on or independent of the patient's age. The property of independence/dependence describes the state of the relationship between events.*

Specifically, at its most fundamental level, independence describes a relationship between two events that is characterized by the fact that the occurrence of one of these events provides no information about the occurrence of the other. An observer who notes the occurrence of one event learns nothing about the occurrence of the second event if the two events are independent. Consider the thought process of a doctor who is examining a patient who may or may not be suffering from a urinary tract infection. During his examination, the doctor may notice the patient's hair color. However, the observation that the patient's hair is black does not influence the likelihood that the patient is suffering from a bladder infection. Hair color is uninformative about the occurrence of the infection, and knowledge of the patient's hair color does not help the doctor one way or other. We say that the two events of hair color and the presence of a urinary tract infection are independent of each another [1].

If E_1 and E_2 are independent, we say that E_1 is independent of E_2, or $E_1 \perp E_2$. If $E_1 \perp E_2$, then $\boldsymbol{P}[E_1 \cap E_2] = \boldsymbol{P}[E_1]\boldsymbol{P}[E_2]$. When the events are independent, then the probability of their joint occurrence need not be zero, but it is simply the product of the probabilities of the independent events.

As a simple example, consider the action of flipping a fair coin twice. We are interested in computing the probability of a head on the first toss and a head on the second toss. Let E_1 be the event of a head on the first toss, and E_2 the probability of a head on a second toss. Then E_1 and E_2 are not mutually exclusive. However, in this circumstance, the events E_1 and E_2 are independent. We can therefore, compute

* Even when words and expressions such as sovereign, autonomous, self-determination, or self-rule are used to describe independence, there is a relationship implied, for example, "sovereign from what?".

$$\boldsymbol{P}[E_1 \cap E_2] = \boldsymbol{P}[E_1]\boldsymbol{P}[E_2] = \left(\frac{1}{2}\right)\left(\frac{1}{2}\right) = \frac{1}{4}.$$

The computation of the joint event probability when the events are independent is typically very simple. In fact, one of the fortunate consequences of selecting simple random samples is that this procedure produces subjects whose experiences are independent of each other. This independence in turn permits the computation of the joint occurrences of events as the product of the individual probabilities. This simple assumption is the bedrock on which the use of commonly used statistical estimators (e.g., means, variances, and incidence rates) are computed.

3.4.5 Dependence

We have seen that when the events are either mutually exclusive or independent, the computation of the probability of their joint occurrence is relatively simple. This is not the case when the events are dependent. However, the dependent circumstance will be the most useful for us in many clinical research monitoring circumstances.

Dependent relationships can be complicated, requiring thoughtful consideration. When two events are dependent, the observer can gain useful knowledge about the possibility that the second event occurred by knowing the first event's occurrence status. Thus, dependent relationships can be very informative—however, the observer must understand the nature of the dependency. Specifically, she must know how to apply her knowledge of the first event's occurrence to update, re-evaluate, and thereby improve her assessment of the likelihood of the occurrence of the second event.

Although physicians in their day-to-day practice may not formally think of events as being dependent, we nevertheless learn to link events in helpful ways. For example, a patient admitted to an emergency room complaining of chest pain will undergo a diagnostic evaluation that will provide information about the likelihood that the patient has suffered a myocardial infarction (MI). This information includes a complete history of the symptoms of the chest pain (e.g., where is the pain located? Is the discomfort a pressure sensation? Is there associated pain in the jaw or the arms? Was there sweating with the discomfort? Was there any nausea or vomiting associated with the episode?) This is followed by a complete medical and family history, leading to a thorough physical exam and evaluation of the patient's blood assays, electrocardiogram, and other cardiac diagnostic tools.

Each of these procedures is designed to reveal useful information about the cause of the patient's chest pain and, based on these evaluations, the treating physician will come to a conclusion and make treatment recommendations. If the diagnostic workup of the patient reveals her to be a 20-year-old female whose chest pain (associated with contusions, soft tissue pain, and swelling) occurred shortly after falling from a horse onto a fence, then the likelihood that the cause of the chest pain is a heart attack is dramatically reduced. Of course, substernal crushing chest pressure–pain in a 59-year-old obese male with a long history of cigarette smoking and hyperlipidemia who has sustained a heart attack in the past and who currently has ST-T segment elevations on his electrocardiogram is a set of circumstances that is easily recognized as predictive of a heart attack. In each case, the events that the patient experienced before and during the emergency room visit were not independent of whether the patient was experiencing a heart attack. In fact, these events were evaluated precisely because they shed important light on the likelihood of a heart attack.* The dependence between the diagnostic findings and the assessment of the likelihood of a heart attack was used to advantage by permitting the findings of the diagnostic testing to change the emergency room doctor's assessment of the likelihood that the patient was having an MI.

Conditional probability is most useful because it is commonly difficult to articulate the nature of the dependency persuasively and completely using other quantitative approaches. Consider, for example, the relationship between health care access and cultural background. It has been well established that some cultures in the United States visit physicians and healthcare providers more commonly, receive prescriptions at a greater frequency, and are more likely to receive prenatal care than others. However, the precise nature of the connection is unknown. There is no formula which accurately and reliably depicts the nature of the relationship, a relationship that remains after adjusting for the easily anticipated sociodemographic predictors of health care access (e.g., age, gender, and the presence of comorbidity. Thus, the nature of the relationship is beyond the ability of sociology to completely discern with modern statistical tools, although its presence is undeniable. By developing the probability of health care access given the person's cultural background, conditional probability allows a precise estimate of the magnitude of the dependency without having to elucidate the dependency's nature.

We may specifically denote the probability of an event A when the event B has occurred as $\boldsymbol{P}[A \mid B]$. This probability may be computed as

$$\boldsymbol{P}[A \mid B] = \frac{\boldsymbol{P}[A \cap B]}{\boldsymbol{P}[B]}. \tag{3.1}$$

The motivation for this formula may be most visible demonstrated graphically (Figure 3.3).

* On the other hand, part of the diagnostic workup of a patient with a possible MI does not include interrogating the patient about their rate of fingernail growth. Rapid fingernail growth provides no useful information about the occurrence of a heart attack, and we say that rapid fingernail growth and the occurrence of an MI are independent events.

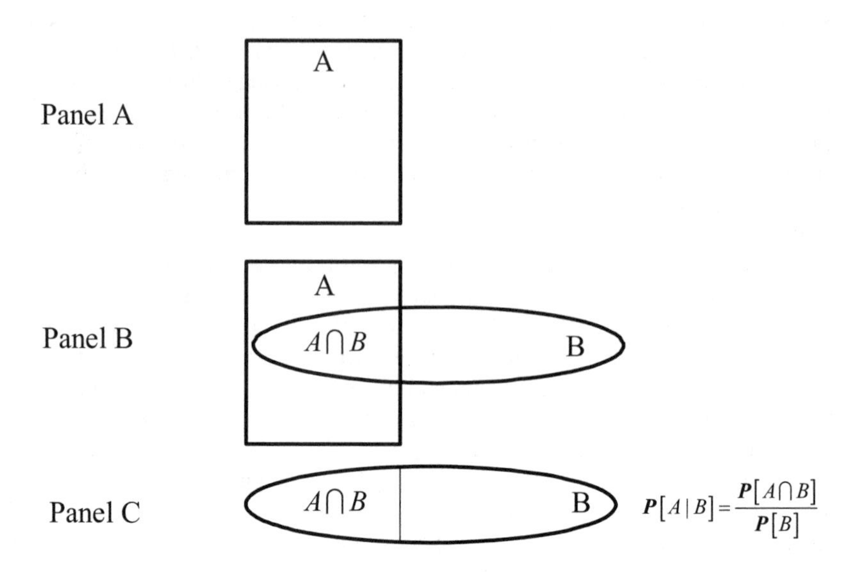

Figure 3.3. Development of conditional probability.

Consider the following example of a conditional probability. An investigator is monitoring a clinical study that is based on the value of the test statistic. The test statistic includes the number of patients in the research effort, and is denoted as T_n. The investigator knows that $\boldsymbol{P}\left[T_{300} > 3\right] = 0.001$, signifying an infrequent event. As the trial progresses, he learns that the value of the test statistic when the number of patients in the trial is 200, T_{200}, is greater than 3, and is told by the biostatistician that $\boldsymbol{P}\left[T_{200} > 3\right] = 0.025$ In addition, the investigator knows that the probability that both $T_{300} > 3$ and $T_{200} > 3$ is 0.013. He is interested in $\boldsymbol{P}\left[T_{300} > 3 \mid T_{200} > 3\right]$. From this information he can compute

$$\boldsymbol{P}\left[T_{300} > 3 \mid T_{200} > 3\right] = \frac{\boldsymbol{P}\left[T_{300} > 3 \cap T_{200} > 3\right]}{\boldsymbol{P}\left[T_{200} > 3\right]} = \frac{0.013}{0.025} = 0.52.$$

Without knowing the size of the test statistic when 200 patients were in the trial, the unconditional probability that the test statistic is greater than three when 300 patients are in the study is 0.001. This is a small number, representing an unlikely event. However, knowing the value of the test statistic at the intermediate point when 200 patients were enrolled in the study increases the probability that $T_{300} > 3$ to 0.52, a 520-fold increase. Knowledge of the value of the test statistic for $n = 200$ informed the process.

Examining the formula for conditional probability, we may write, in the case of dependent events, that $P[A \cap B] = P[A \mid B]P[B]$. As was the case with the scenario of independence, the probability of the joint occurrence of events is the product of probabilities. However, when the events are dependent, one of the probabilities to be used in this product is a conditional probability.

Thus, when faced with computing the probability of the joint occurrence of events *A* and *B*, we just decide if the events of interest are mutually exclusive, independent, or dependent. If the events are mutually exclusive, then $P[A \cap B] = 0$. If the events are independent, then $P[A \cap B] = P[A]P[B]$. Finally, if the events are dependent, then $P[A \cap B] = P[A \mid B]P[B] = P[B \mid A]P[A]$ (Figure 3.4).

Event	Probability
A and B are mutually exclusive	➡ $P[A \cap B] = 0.$
A and B are independent	➡ $P[A \cap B] = P[A]P[B].$
A and B are dependent	➡ $P[A \cap B] = P[A \mid B]P[B]$ $= P[B \mid A]P[A].$

Figure 3.4. Computing the probability of joint events.

3.4.6 Unions of Events

The implementation of statistical monitoring in clinical trials requires us to identify and combine simple events into more complicated events of interest. The intersection is one such tool. A second device of interest is the union of events.

The union of two events signifies that at least one of them occurs. We write the possibility that either event *A* occurs or event *B* occurs as $A \cup B$, commonly stated as "*A* union *B*", or "*A* and/or *B*".

Computing the probability of this event follows directly from examination of the possible ways that this can occur (Figure 3.5).

From Panel A of Figure 3.5, the events *A* and *B* are mutually exclusive. In this case the probability of the occurrence of either the event *A* or the event *B* is simply the sum of the probabilities: $P[A \cup B] = P[A] + P[B]$. However, in Panel B where the two events *A* and *B* are not disjoint, the simple addition of the probabilities of the events is inappropriate. This is because both event *A* and event *B* include the event $A \cap B$. Thus, in the case of nondisjoint events, the simple addition of $\boldsymbol{P}[A]$ and $\boldsymbol{P}[B]$ would include the $\boldsymbol{P}[A \cap B]$ twice; we must remove one of these redundant components. Thus $\boldsymbol{P}[A \cup B] = \boldsymbol{P}[A] + \boldsymbol{P}[B] - \boldsymbol{P}[A \cap B]$.

The consideration of event unions is essential in the monitoring of clinical research. One common scenario occurs when the monitoring investigator is concerned about the occurrence of either a beneficial effect or a harmful effect. Consider the example in which a clinical study's protocol contains the prospective declaration that its results will be monitored 20% into the trial's total duration of follow-up.

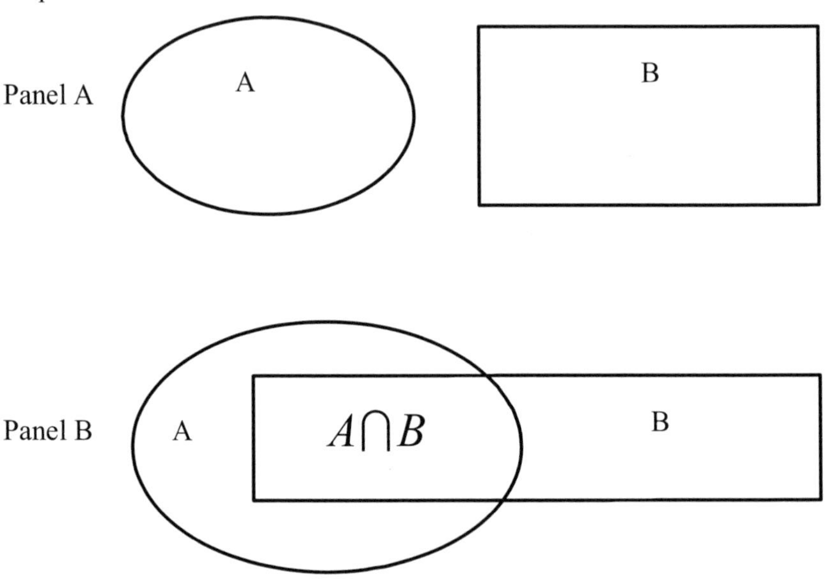

Figure 3.5. Examination of the union of two events A and B.

The test statistic at this point is denoted $T_{20\%}$. The evaluation of interest is the effect of the intervention on the change in National Institutes of Health Stroke Score (NIHSS). It is determined that $T_{20\%} \geq 3.5$ would be an overwhelming positive effect. Alternatively, $T_{20\%} \leq -2.5$ would reflect important harm. Thus, the investigator is interested in $\boldsymbol{P}[T_{20\%} \geq 3.5 \cup T_{20\%} \leq -2.5]$. Because these two events are mutually exclusive, then this probability can be decomposed into the sum of two probabilities, $\boldsymbol{P}[T_{20\%} \geq 3.5 \cup T_{20\%} \leq -2.5] = \boldsymbol{P}[T_{20\%} \geq 3.5] + \boldsymbol{P}[T_{20\%} \leq -2.5]$.

3.4.7 Unions and Intersections in Monitoring

The preceding elementary example that involved monitoring clinical events involved mutually exclusive outcomes. However, a circumstance in which the events of interest are not mutually exclusive is readily available. Suppose the same investigator, in addition to the concern about the effect of the intervention on improvement in the NIHSS, is also concerned about a harmful effect on the rate of intracerebral bleeding events. In this case, we have another test statistic, $S_{20\%}$. that reflects the occurrence of intracerebral bleeds when 20% of the total time of the trial has elapsed. The investigator would consider terminating the study if $S_{20\%} < -1.5$.

In this case, the event to stop the trial is the more complicated union of $T_{20\%} \geq 3.5$ and/or, $T_{20\%} \leq -2.5$ and/or $S_{20\%} \leq -1.5$. The region of interest can be readily identified (Figure 3.6). We can write from an observation of this figure that

$$\begin{aligned}&\boldsymbol{P}\left[T_{20\%} \geq 3.5 \cup T_{20\%} \leq 2.5 \cup S_{20\%} < -1.5\right] \\ &= \boldsymbol{P}\left[T_{20\%} \geq 3.5\right] + \boldsymbol{P}\left[T_{20\%} \leq 2.5\right] + \boldsymbol{P}\left[S_{20\%} < -1.5\right] \\ &-\boldsymbol{P}\left[T_{20\%} \geq 3.5 \cap S_{20\%} < -1.5\right] - \boldsymbol{P}\left[T_{20\%} \leq -2.5 \cap S_{20\%} < -1.5\right].\end{aligned} \quad (3.2)$$

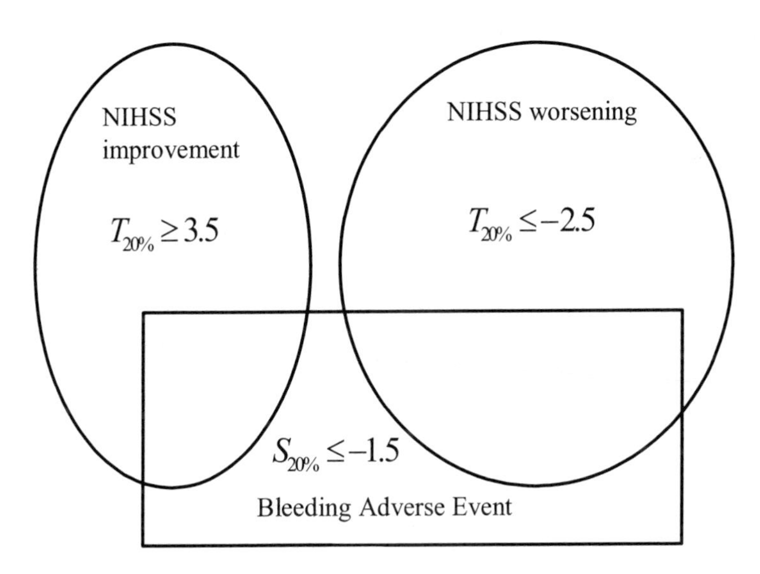

Figure 3.6. Union of efficacy and safety events in a clinical study.

Although the probabilities of each of the individual events (commonly called marginal probabilities) $\boldsymbol{P}\left[T_{20\%} \geq 3.5\right]$, $\boldsymbol{P}\left[T_{20\%} < -2.5\right]$, and $\boldsymbol{P}\left[S_{20\%} < -1.5\right]$, can be more easily identified, the two joint probabilities $\boldsymbol{P}\left[T_{20\%} \geq 3.5 \cap S_{20\%} < -1.5\right]$ and $\boldsymbol{P}\left[T_{20\%} \leq -2.5 \cap S_{20\%} < -1.5\right]$ require additional work. At first examination, the evaluation of these probabilities can appear to be difficult. However, their values may be more accessible by writing them as the probability of joint events, for which we can apply our previous discussion of conditional probability. For example, we may write

$$\boldsymbol{P}\left[T_{20\%} \geq 3.5 \cap S_{20\%} < -1.5\right] = \boldsymbol{P}\left[S_{20\%} < -1.5 \mid T_{20\%} \geq 3.5\right] \boldsymbol{P}\left[T_{20\%} \geq 3.5\right],$$

and

$$\boldsymbol{P}\left[T_{20\%} \leq -2.5 \cap S_{20\%} < -1.5\right] = \boldsymbol{P}\left[S_{20\%} < -1.5 \mid T_{20\%} \leq -2.5\right] \boldsymbol{P}\left[T_{20\%} \leq -2.5\right].$$

Substituting these results for the joint probabilities into expression (3.2), we have

$$\begin{aligned}
&\boldsymbol{P}\left[T_{20\%} \geq 3.5 \cup T_{20\%} \leq 2.5 \cup S_{20\%} < -1.5\right] \\
&= \boldsymbol{P}\left[T_{20\%} \geq 3.5\right] + \boldsymbol{P}\left[T_{20\%} \leq 2.5\right] + \boldsymbol{P}\left[S_{20\%} < -1.5\right] \\
&-\boldsymbol{P}\left[S_{20\%} < -1.5 \mid T_{20\%} \geq 3.5\right] \boldsymbol{P}\left[T_{20\%} \geq 3.5\right] \\
&-\boldsymbol{P}\left[S_{20\%} < -1.5 \mid T_{20\%} \leq -2.5\right] \boldsymbol{P}\left[T_{20\%} \leq -2.5\right].
\end{aligned}$$

The joint probability of the threshold event has been disassembled into marginal probabilities and conditional probabilities. The advantage offered by this formulation is that the conditional probabilities are usually more easily estimated by clinicians.

3.4.8 Complements

One final useful feature in constructing events of interest is the use of complements. Any element that is not in the set A is in the complement of A. The complement of a set is its opposite. The complement of the set A is the set A^c, termed "A complement" or "Not A". In general $\boldsymbol{P}\left[A^c\right] = 1 - \boldsymbol{P}\left[A\right]$.

The consideration of complements of clinical sets of interest can be particularly valuable. As an illustration, consider the example of a clinical trial that is designed to determine the effect of therapy on patients who have diabetes mellitus. The intervention of the study is such that the investigators consider developing a monitoring rule that is subgroup-dependent, that is, a different threshold for discontinuing the trial is developed for different subgroups of patients.* In this circumstance a monitoring rule is sought for patients with profound disease (in this case, patients who suffer from diabetic retinopathy and diabetic nephropathy), that is different from the monitoring rule that is developed for the remaining patients.

However, who are the remaining patients? If we define the set R as patients with diabetic retinopathy and N as the set of patients with diabetic nephropathy, then $R \cap N$ signifies patients who have both sequela. The complement of this set, $(R \cap N)^c$ is interesting (Figure 3.7). The complement of $R \cap N$ consists of diabetes patients who have neither diabetic retinopathy nor diabetic nephropathy, or only one. The complement of the intersection is the union of the complement of R (i.e., any diabetic patient with neither retinopathy nor nephropathy or isolated reti-

* This must be defined very carefully, because drawing conclusions from subgroup analyzes can be hazardous. See [1] for a discussion of the difficulties posed by undisciplined subgroup analyzes and more references.

nopathy), and the complement of N (i.e., any diabetic patient with neither retinopathy nor nephropathy, or isolated nephropathy).*

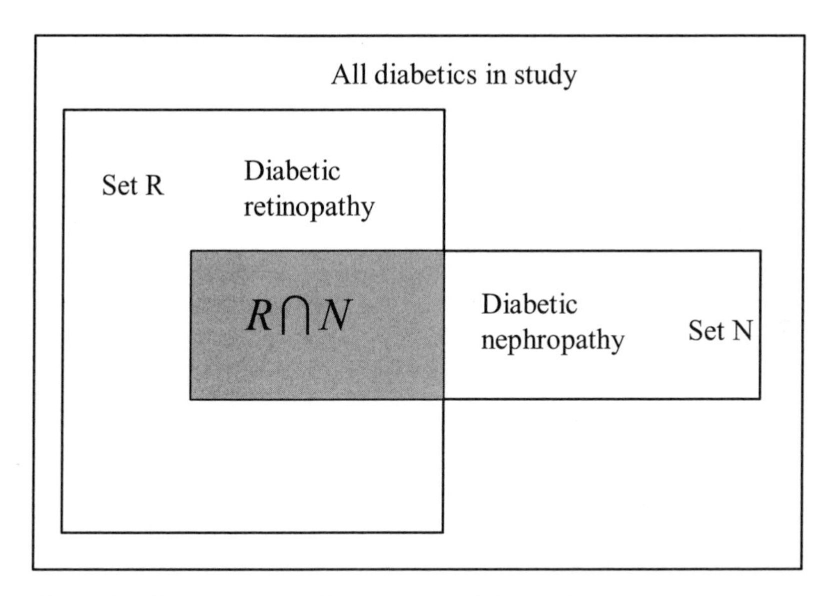

Figure 3.7. Set complements for monitoring diabetes trial.

3.5 Investigator Perplexity

The ability of probability to operate on collections or sets of objects provides important flexibility for its application. We have already seen some important uses of this set function. However, we will have to broaden the way that this set function operates if we are to use it to full advantage in its application to monitoring clinical research.

For example, consider the results of tosses of two flips of a fair coin, whose outcomes are $\{H\,H\}$, $\{H\,T\}$, $\{T\,H\}$, and $\{H\,T\}$. Each of these four outcomes is discrete and has probability assigned to it (Figure 3.8). In many circumstances in which the outcomes are individualized and discrete, we can assign probability to a point. This is the simplest approach to the application of probability. However, this intuitive procedure, although satisfactory in many settings, will not work when there are too many outcomes. There are common circumstances in healthcare research in general, and statistical monitoring of clinical research in particular, where there are simply too many possibilities for the procedure of assigning a positive probability to each will work.

For example, consider an investigator who is monitoring a clinical trial that is evaluating the effect of an intervention. When 25% of the follow-up time of the study is completed, he is told that the value of the test statistic at this interim

* DeMorgan's law states that $(A \cup B)^c = A^c \cap B^c$ and $(A \cap B)^c = A^c \cup B^c$.

point is $T_{25\%} = 2.08$. A reasonable question for him to ask is, "How likely is the value of that test statistic under the null hypothesis of no therapy effect?"

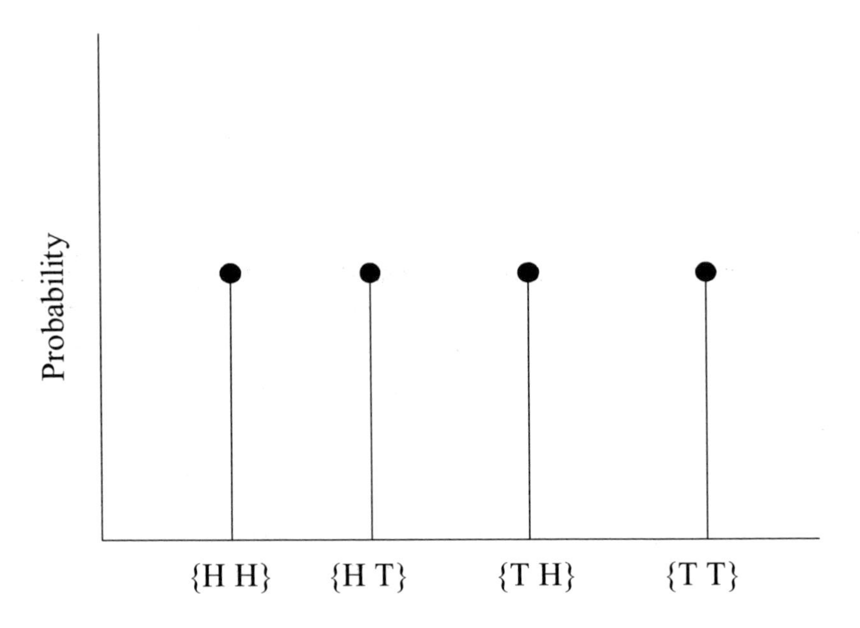

Figure 3.8. Probability as point mass.

This is a clear and reasonable question. The investigator has an interim outcome, and simply desires to learn the probability of that outcome. In order to answer this question, we need to compute the probability of all outcomes so that the relative frequency of the value 2.08 can be calculated. However, every number is a possible outcome, and any way that we try to assign positive probability to each number in the set of all numbers produces a total probability that is greater than one. If the investigator attempts to make the problem easier by restricting the possible values of the test statistic to only positive numbers, the answer is the same. This attempt to simplify the problem by reducing the number of possibilities by half did not help. In fact, any way that we try to assign nonzero probability to all of the numbers in an interval will fail.

This result understandably perplexes many investigators. The paradigm that we have thus far used to compute probability has permitted probability to be assigned for each set. Probability can be assigned in this way so that the events that occur from actions that produce a large number of finite events can be computed. There are even cases where positive probability can be assigned to an infinite number of events that may result from an action. So it seems peculiar that some actions which produce an infinite number of events can have positive probability assigned to them, and others cannot. What is getting in our way is our ability (or inability) to count.

3.5.1 Countably versus Uncountably Many

Positive probability can be assigned when the events are denumerable or countable. This simply means that there is a way to either count them all, or alternatively, to set up a procedure such that, if we had enough time, we could count them all.

In the previous example of two flips of a coin, we can clearly count the four outcomes $\{H\,H\}$, $\{H\,T\}$, $\{T\,H\}$, and $\{H\,T\}$. An example of an action that produces an infinite number of possible outcomes is the number of radioactive particles that are delivered to a tumor site during a treatment period for a patient's liver cancer. The result of this action produces the set of non-negative integers 0, 1, 2, 3, … . There are an infinite number of them, but if we had enough time, we could count them all. By simply going in order, we would not miss any of them. It would just take us forever[2].*

However, some actions produce many more outcomes. An example is an experiment that randomly produces a number between zero and one. How can we count all of these numbers, each one representing a different outcome? We start the counting process with zero, but what follows that? Any choice that we make for the second number will miss an infinite number of others. Not only can we not count them all, we cannot set up a mechanism of counting them that will enumerate all of the numbers in this [0, 1] interval. We say that this set of numbers on the [0, 1] interval is *nondenumerable,* or *dense,* and that the [0,1] interval contains *uncountably many* elements. Density is a property of any interval on the real number line. It is this property of density (i.e., the fact that there are too many numbers to know how to count) that foils our attempts to assign positive probability to each and every number lying within the interval.

We must be careful here. The problem that has arisen is not with the concept of probability. It is the measurement of probability as point mass that must be set aside. In this new paradigm, the probability of a given individual point has no meaning. We therefore replace it with the idea of assigning probability not to a point, but to an interval (Figure 3.9).

---

* The typical probability model for this action is the Poisson model. If Y is the number of particles that are detected from time (0, t), then $\boldsymbol{P}(Y=k)=\frac{(\lambda t)^k}{k!}e^{-\lambda t}$, where λ is the intensity of particle arrival. This probability model has seen widespread use, including predicting the number of arrivals to an emergency room, the likelihood of judicial vacancies on the United States Supreme Court, and the frequency of kicks that soldiers received from horses in the 19th century Prussian Army!

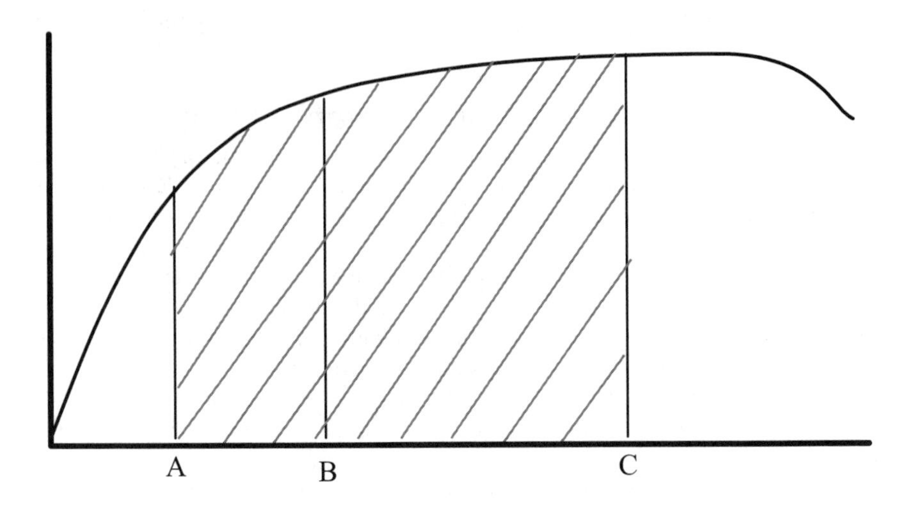

Figure 3.9. Probability as area. The individual points on the x-axis A, B, and C have no probability attached to them, but the interval from A to C has probability as measured by the area under the curve (cross-hatched).

A simplified explanation would be that, because all of the possible events (i.e., points in an interval) are not denumerable, we cannot assign a positive probability to each one. We therefore, combine adjacent points into intervals. This collection of intervals is denumerable, and we can assign or compute a positive probability to any one of them. This probability is assigned as area.

Returning to the plight of our investigator who was left holding the interim result of $T_{25\%}= 2.08$, our answer to the question of, "What is the probability of this occurrence," must be, "zero." However, we can compute a more helpful solution by changing the relevant event from a single point to that of an interval. For example, we can compute the probability that the test statistic at this interim point in the study lies in the interval from 2.08 to infinity (i.e., is more extreme than 2.08).

Thus, when outcomes of actions are all of the points in an interval of the real number line, we replace the idea of the probability of a point with the idea of probability as area.*

3.5.2 Applicability to Monitoring Clinical Studies

The previous conversation that led us to set aside the notion of only using point mass for probability has two important implications for the investigator interested in the statistical monitoring of clinical research.

The first implication is that he will be in the same position as the investigator who had the interim value of the test statistic in the illustration provided in the

* The idea of combining point mass probability and the probability as area is an exciting one (at least, to probabilists!). It has a direct application to the Bayesian approach to the statistical monitoring of clinical results, and will be discussed in Chapter 8).

previous subsection. A decision-making process may be triggered by the interim value of a test statistic. However, as we have seen, the probability of any one value of the test statistic is zero. Prospective knowledge of the difficulty with computing the probability of a single value of the test statistic motivates the investigator to an a priori focus on regions or intervals of possible test statistics that are of clinical relevance to him.

Secondly, this problem is compounded by the fact that, just as there are uncountably many numbers on the real line, there are also uncountably many paths that a test statistic can follow over time. We can now anticipate that the answer to the question, "What is the probability of the path that my test statistic took" will also be "zero." This will motivate us to focus not on a single individual path, but on collections of paths.

3.6 The Normal Distribution

3.6.1 Introduction

Although there are many probability distributions that have been discovered over the past 300 years, the most frequently used one is the "bell-shaped curve." (Figure 3.10).

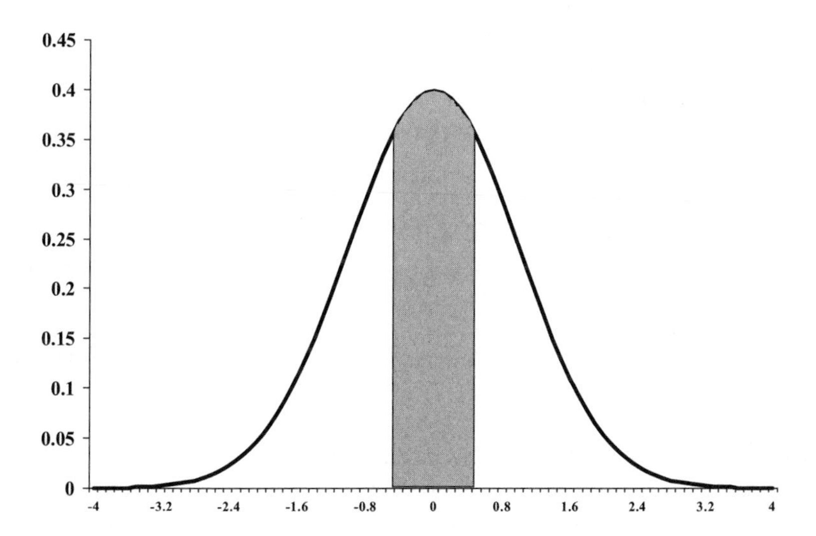

Figure 3.10. Probability of an interval is provided as an area under the normal distribution.

Formally known as the Gaussian distribution,* it is the probability distribution that is commonly used in the statistical monitoring of clinical trials. The events whose

* Named for Karl Friedrich Gauss (1777–1855), a calculating prodigy who lived in the 18th century. His contributions to mathematics were numerous, include the development of the

probability it provides are intervals on the real number line. The likelihood of any interval on that real line is the area under the curve depicted in Figure 3.12.

The formula for the curve is complicated. If z is a point on the real line than $f(z)$ is computed as

$$f(z) = \frac{1}{\sqrt{2\pi}} e^{-z^2/2}. \tag{3.3}$$

The complexity of this formula begs the question of how could an expression that is so esoteric be so useful. There are two reasons for its ubiquity. The first is that, occasionally, an action is carried out whose outcomes have the precise probability that is the area under equation (3.3).

The second, most common reason for its omnipresence in applied probability focuses on the nature of outcomes from actions. These events are commonly composed of the resultant of many influences. These influences are small and directional. Some of these influences produce positive effects whereas others have negative effects. The sum of these myriad small effects produces a resultant outcome that follows the normal distribution. Although the mathematics of this are quite precise,* the implications are remarkable. In fact, the widespread use of this result is the motivation for the nickname of "normal distribution" for the bell-shaped curve.

An example of this was provided in Chapter One in which the movement of pollen grain was produced by the sum of positive and negative effects. In fact, a comparison of the rightmost side of expression (1.1) in Chapter One with expression (3.3) above demonstrates that the formulas are identical. We will rely on this correspondence as we discuss the particular statistical monitoring procedures in subsequent chapters.

3.6.2 Using the Normal Distribution

An important advantage in using the normal distribution is the relative ease of producing probabilities from it. Although the actual production of probabilities requires one to use a table (Appendix E), fortunately we need use only one table to produce probabilities for the many different normal distributions. This is due to the ease of transformation of normal random variables and the notion of symmetry.

3.6.2.1 Simplifying Transformations

The normal distribution is not just a single distribution, but is instead a family of them. There are an infinite number of normal distributions, each one characterized by its unique mean μ and variance σ^2. The mean provides the location of the distribution, and the variance characterizes the dispersion of the variable around that mean (Figure 3.11).

---

normal distribution, contributions to calculus theory, and the discovery of the "least squares" approach to model building.

* The central limit theorem governs this use of the normal distribution.

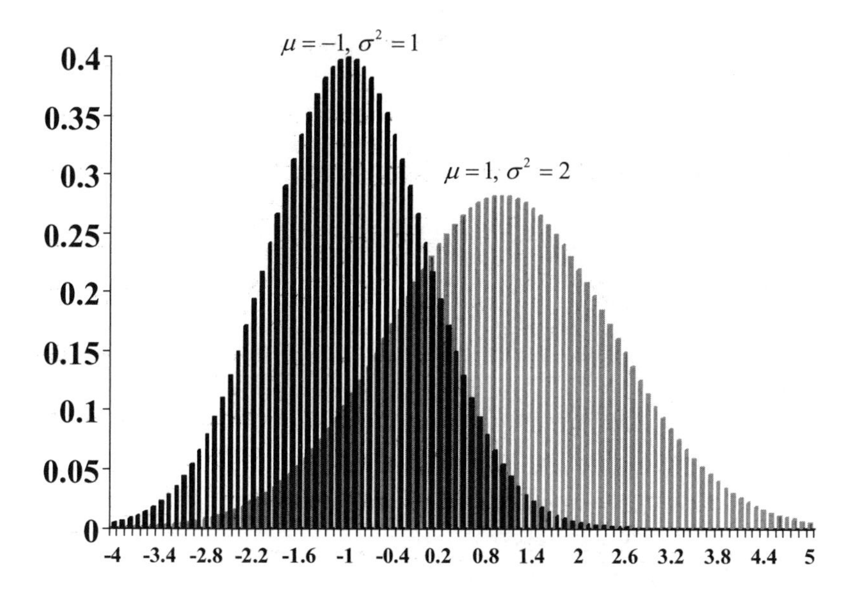

Figure 3.11. Different locations and dispersions of two normal distributions.

Figure 3.11 provides two examples, or members from the family of normal distributions, their locations and dispersal or "spread" governed by the mean and variance of each.

The fact that the location and the shape of a normally distributed variable is governed by its mean and variance suggests that the investigator must incorporate these two parameters into any computation of the probabilities using these distributions. Because there are an infinite number of combinations of these parameters, a first impression is that these computations can rapidly become unwieldy. Fortunately, this complexity is removed by the fact that each member of the family of normal distributions can be related or transformed to another normal distribution. This observation produces the simplifying principle that any normal variable can be related or reduced to a single normal distribution. This single normal distribution has a mean of zero and a variance of one, and is known as the *standard normal distribution*.

Therefore, if X follows a normal distribution with mean μ and variance σ^2, then X, and any event involving X can be transformed to a normally distributed variable with mean 0 and variance 1. Specifically, if X follows a normal distribution with mean μ and variance σ^2, then $(X-u)/\sigma$ follows a normal distribution with mean 0 and variance 1.

Because every normally distributed random variable can be converted to a standard normal distribution, we require only a single set of tables that provide probabilities from the normal distribution. For example, we can use this fact to compute the probability that each of the normally distributed variables that appears in Figure 3.11 is less than the value 0.50. For the first, if X follows a normal distribution with mean 1 and variance 2, then

$$\boldsymbol{P}[X < 0.50] = \boldsymbol{P}[X - 1 < 0.50 - 1] = \boldsymbol{P}\left[\frac{X-1}{\sqrt{2}} < \frac{0.50-1}{\sqrt{2}}\right]$$
$$= \boldsymbol{P}\left[Z < \frac{0.50-1}{\sqrt{2}}\right] = \boldsymbol{P}[Z < -0.353] = 0.362.$$

For the second case, we compute

$$\boldsymbol{P}[X < 0.50] = \boldsymbol{P}\left[Z < \frac{0.50+1}{1}\right] = \boldsymbol{P}[Z < 1.50] = 0.933.$$

We will take advantage of this tool repeatedly in our work in computing probabilities involving monitoring procedures.

3.6.2.2 Symmetry

A second useful characteristic of the normal distribution is its property of symmetry. Examination of the curves in Figures 3.10 and 3.11 reveal that the shape of the normal distribution is a mirror image of itself when divided at the mean. This produces some useful computation simplifications. Begin with a variable Z that follows the normal distribution. We know that the $\boldsymbol{P}[Z = 0] = 0$ from the concept of probability as area. We also know from use of the property of symmetry that $\boldsymbol{P}[Z < 0] = \boldsymbol{P}[Z > 0]$. Because the sum of these probabilities must equal one, we see that $\boldsymbol{P}[Z < 0] = \boldsymbol{P}[Z > 0] = \frac{1}{2}$.

In fact, the use of symmetry produces the relationship that for any value z, $\boldsymbol{P}[Z < -z] = \boldsymbol{P}[Z > z]$ (Figure 3.12). This style of computation is frequently utilized. If we define $F_Z(z) = \boldsymbol{P}[Z \leq z]$, then we know that $F_Z(-z) = 1 - F_Z(z)$. This equality is used to compute the p-values for two-sided testing from the normal distribution. Let TS be the test statistic that is produced from a statistical hypothesis test, and $|TS|$ be its absolute value. Then, if TS follows a standard normal distribution,

$$p-value = \boldsymbol{P}\left[Z \leq -|TS|\right] + \boldsymbol{P}\left[Z \geq |TS|\right] = 2\boldsymbol{P}\left[Z \geq |TS|\right].$$

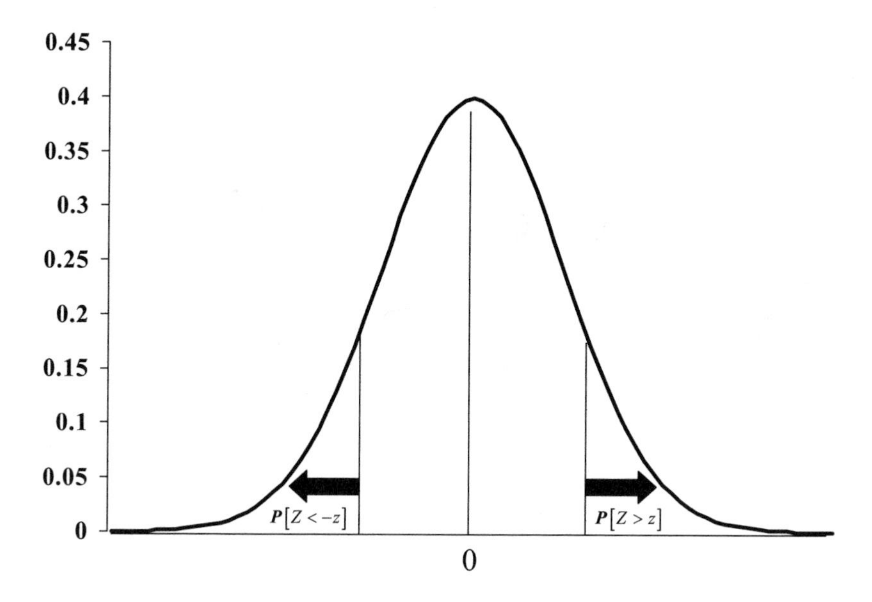

Figure 3.12. Symmetry of the normal distribution.

3.7 Probability and Monitoring Procedures

The value of probability for us is to provide a metric to assess the relative likelihood of various outcomes of a clinical study that has not yet been completed. We are now in a position to begin to see how this would work.

An investigator who is responsible for monitoring a study has the value of the test statistic S_0 at the study's inception (time 0). He also has the value of the test statistic S_t at some interim time t. With no intermediate values of the test statistic, this investigator cannot know the exact path that the test statistic took as it moved from S_0, to S_t. He has only these two values (Figure 3.13).

The investigator would like to know how extreme the value of the test statistic S_t is at time t. Another way to phrase this inquiry is, how extreme is the path that produced the test statistic S_t? A little thought reveals that there are many paths that the test statistic could have taken to arrive at the value S_t at time t. In fact, there are uncountably many paths that the test statistic could have followed. If the investigator is interested in computing the relative likelihood of a path that produced the test statistic S_t, then the counting problem he faces is immense.

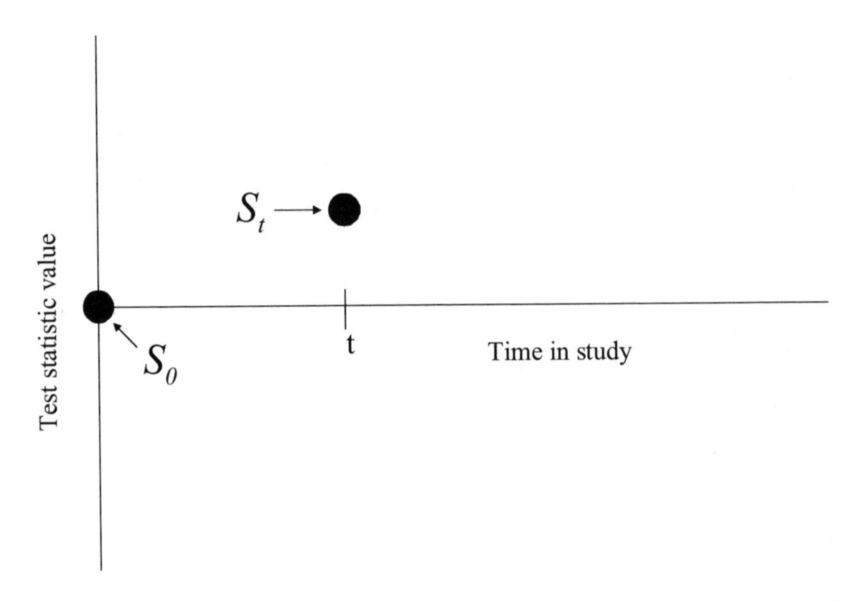

Figure 3.13. Measurement of test statistic in a clinical study at the interim time point t.

However, the problem is solvable if we move away from consideration of paths that all wind up at the same point, and instead move to the idea of computing the relative frequency of test statistics that lie in an interval of interest. An example would be the region of test statistics that are at least as extreme as the observed test statistic. If the investigator can identify the collection of test statistic paths that each produce a test statistic falling within the interval of interest, we are closer to our goal, because we can use the probability as area concept to produce the probability of that interval. By computing the probability of that interval, we have learned the probability of the collection of paths that produce test results in that interval. This is one use of statistical monitoring in clinical research (Figure 3.14). In this case, the likelihood of more extreme values of the test statistic is computed by applying the normal distribution to the interval of more extreme values.

This rather sophisticated application of probability must be further motivated, which is the topic of Chapter 4.

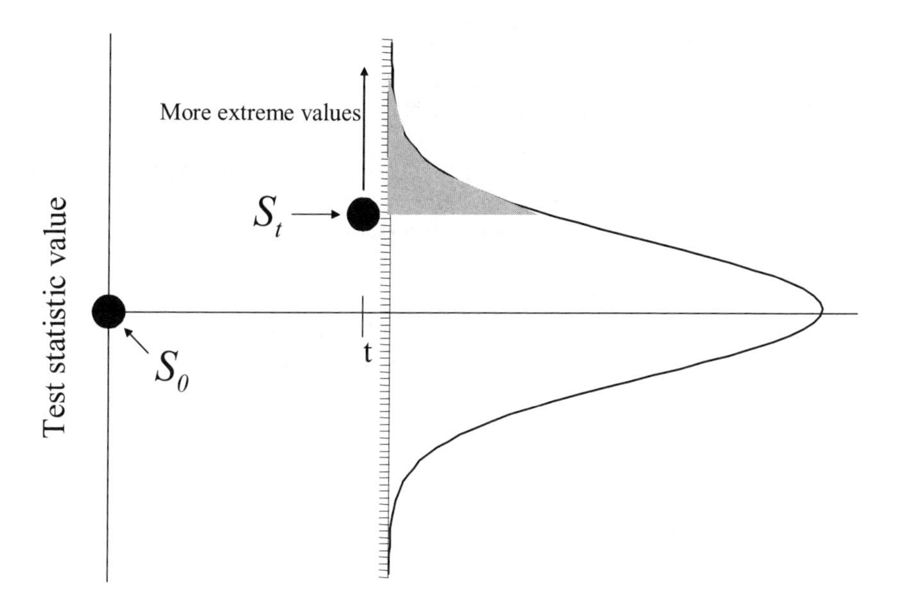

Figure 3.14. Measurement of test statistic in a clinical study at the interim time point *t* with probability assessment. The shaded region is the probability of paths that produce more extreme values of the test statistic at time *t*.

Problems

1. An investigator is studying the effect of a new therapy to reduce the extent of the neurologic damage produced by stroke. The only known serious adverse event produced by the therapy is intracranial hemorrhage (ICH). The probability that any one patient suffers from an ICH 0.14. Your statistician tells you that, if the occurrence of an ICH in one patient does not inform you one way or another about the occurrence of an ICH in another patient, then the probability that there are *k* patients who suffer in ICH in this case series of 42 patients is

$$\boldsymbol{P}(X=k)=\frac{42!}{k!(42-k)!}(0.14)^k(0.86)^{n-k}.$$

Using this formula, compute the following.

a. **P**[exactly one patient in this study suffers an ICH].
b. **P**[no patients suffer an ICH].
c. **P**[at least one in this study who suffers an ICH].
d. **P**[between 5 and 10 patients in this study suffers an ICH].
e. **P**[either less than 5 or greater than 35 patients suffers an ICH].

For the following problems, let T_p be the value of the test statistic for the effect of therapy in a well-designed, well-executed clinical trial when p percent of the follow-up period has elapsed. Thus, T_{50} is the value of the test statistic when 50% of the trial has been completed and T_{100} is the value of the test statistic when 100% of the follow-up period of the trial has elapsed (i.e., when the trial has ended).

2. An investigator is interested in designing a monitoring guideline to help her discontinue a study, based on the value of the test statistic. Let $T_{.20}$ represent the value of the test statistic when discontinuing a study. She would like to compute the likelihood that the test statistic at the conclusion of the study is greater than 1.96, given the value of the test statistic when 20% of the follow-up period has elapsed, T_{20}. She is told that the $\boldsymbol{P}[T_{20} > 1.96] = 0.34$. Show

$$\boldsymbol{P}[T_{100} > 1.96] = 0.34\boldsymbol{P}[T_{100} > 1.96 \mid T_{20} > 1.96] + 0.66\boldsymbol{P}[T_{100} > 1.96 \mid T_{20} < 1.96].$$

3. An investigator is interested in tracking a test statistic's performance over time. He knows that $\boldsymbol{P}[T_{30} \geq 3.5] = 0.005$. Complete the last column in Table 3.1.

Table 3.1. Conditional Probabilities of Reaching a Threshold at T_{30}

X	$\boldsymbol{P}[T_X \geq 3.5]$	$\boldsymbol{P}[T_{30} \geq 3.5 \cap T_X \geq 3.5]$	$\boldsymbol{P}[T_{30} \geq 3.5 \mid T_X \geq 3.5]$
5	0.00010	0.0000030	
7	0.00050	0.0000040	
10	0.00070	0.0000100	
15	0.00090	0.0000900	
20	0.00100	0.0005050	
25	0.00300	0.0025000	

Does information about the distribution of the test statistic when 30% of the information is available depend on the distribution of the test statistic at earlier points in the execution of the trial?

References

1. Moyé L. (2003) Multiple Endpoints in Clinical Trials: Fundamentals for Investigators. New York Springer.
2. Parzen E. (1960) *Modern Probability Theory and its Applications*. New York, Wiley.

4

Issues and Intuition in Path Analysis

4.1 Pondering the Meaning of It All

Monitoring is an ethical requirement of modern clinical research, and its application must balance the need for the study to end when clear evidence demonstrates the direction and magnitude of the treatment effect with the requisite for precise and persuasive estimates of that treatment effect. In addition, difficulties generalizing from a sample to a population are compounded by the new problem of generalizing from only a fraction of the sample to the larger population.

In addition, the investigator is beset by the vexing problem of drawing conclusions from a stream of unpredictable data observed for only a short period of time. Observing a random process for a relatively brief time during which one hopes to see its long-term target presents its own problems, as the following illustration, adapted from [1] demonstrates.

> In a rush to get to the hospital, the university, the airport, or the institute, you get caught in a traffic jam that stretches out for miles in front of you There are two lines of slowly moving traffic, and you slide into one, nudging in behind the car in front of you. As it turns out, I am in the car immediately behind you, and because you chose one lane, I choose the other. There we are, side by side in two different lanes of traffic, sitting still more than we are moving. You have a statistician in your car as a passenger, and nudging him, you point to my car, saying,
>
> "You're an expert in these things. How likely is it that we will pull ahead of Lem?"
>
> Answering with smooth assuredness, he says, "You will definitely pull far ahead of him. Just be patient."
>
> However, as you watch, I slowly pull ahead of you. First I am one car length ahead of you, then two, and then three full car lengths. A few minutes later, I am out of sight. You are befuddled, but your statistician still appears remarkably confident. Turning to him, you remark, "I thought that you said I would pull ahead of him?"
>
> "Yes, I did. And I will say it again. You will pull far ahead."

"Well, just when was that going to happen?"
"You have to wait an infinite amount of time."
It just takes you a moment to compose your answer, "So do you. Get out!"
After your passenger steps out, now himself looking befuddled, you are left alone (still stuck in traffic) to ponder the meaning of it all.

The trends appearing in research data that are collected, analyzed, and then sent to a DMC can appear to be like the relative movement of traffic in the preceding example. First a pattern appears, suggesting one result. This pattern then disappears, replaced by another ephemeral trend. One reason that results from random data are so challenging is because their content can frequently appear to be unreliable. We want to believe what they "say", but if we wait a little longer they "say" something else. Perhaps it is not quite fair to conclude that this incoming data is misleading, but it can be fickle.*

4.2 Generalization Complexities

In Chapter Two, we described the types of misleading conclusions from samples that are the product of sample-to-sample variability or sampling error. Because sampling error is embedded in the data, actually insinuating itself into the study results, a careful and discerning eye is required to separate the signal that reflects a true population effect from the background noise.

The circumstance is not less, but more, perilous when scientists contemplate discontinuing a study early. Consider an investigator who is designing a study of 400 patients in order to examine the effect of a new therapy on the reduction of systolic blood pressure (SBP) in patients with isolated systolic hypertension (ISH). Understanding that any attempt to generalize the results of a 400 patient study to a population of millions of patients might be hazardous, she proceeds carefully. She chooses a sample size that was computed based on her best estimate of (1) the effect of her therapy on SBP, (2) the variability of SBP, and (3) acceptable levels of type I and type II error rates. In addition, she selects her subjects as randomly as possible in order to reduce the problems that she would have in extending her findings.

Finally, she also knows that her assumptions about the study findings might be wrong. For example, the intervention might be far more effective in reducing SBP than she anticipated. Alternatively, a new and unanticipated adverse effect may appear in these subjects that would require her to terminate the study early. In recognition of these possible outcomes, she begins the thought process that would lead to the justifiable early termination of the study.

It is relatively easy to understand the ethical difficulties that would ensue if premature study termination were not possible. However, early termination is not without its own set of difficulties. A principal problem involves generalizability.

* Another point of view is that we are just too impatient. If we could wait an infinite amount of time, the clear solution would be readily apparent.

The investigator's initial plans required her to recruit, randomize, and follow 400 patients in order to provide assurances to the medical and regulatory community that sampling error had been adequately controlled (Figure 4.1 Panel 1). Now, she is contemplating the possibility that she may have to end the study before that goal is achieved. Thus, in the case of early termination, there are two levels of generalization. The first involves extending the results of the analysis that was carried out on only a fraction of the data to the entire sample; the second is to generalize from the sample to the entire population (Figure 4.1Panel 2)

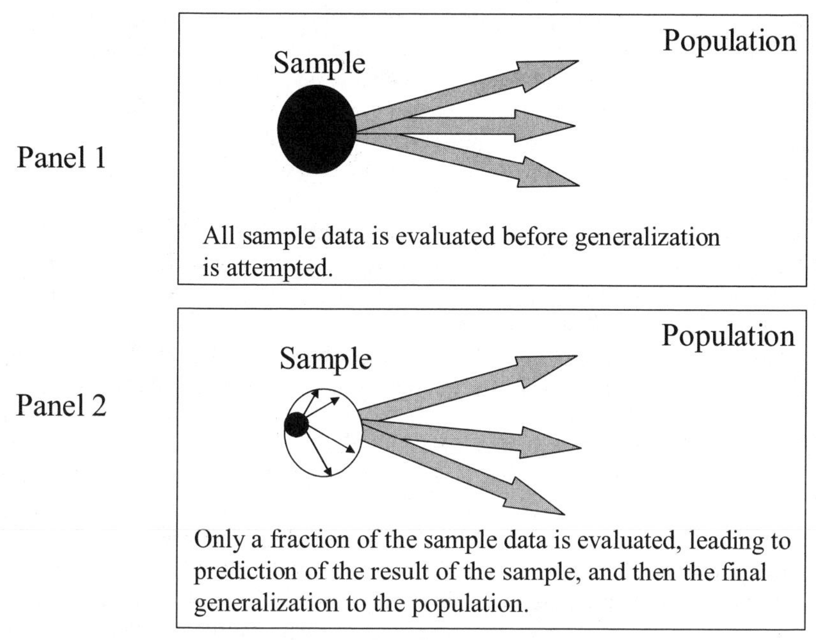

Figure 4.1. Comparing the task of generalizing a sample result to the population (Panel 1), to generalizing from a fraction of the sample to the population (Panel 2).

Thus, in circumstances where there will be early termination of a research effort, there must be especially tight control over the type I error rate if the result is to survive this two-step generalization process. In addition, we must ensure that the estimates of effect sizes are especially trustworthy in order to successfully guide the investigator through these multiple levels of generalization to a clear view of the population findings.

4.3. Why Are Special Tools Necessary?

We have provided ample motivation for the need to monitor clinical studies; however, we have not yet justified the use of any particular procedure to carry out this monitoring. In fact, the investigator who has observed the wealth of tests already in service* may be forgiven for asking the question, "Why is yet one more test necessary?" However, there are three new issues in clinical trial monitoring that the classic hypothesis testing procedures do not directly address. They are (1) repeated testing of data, (2) variability, and (3) the notion of dependency.

4.3.1 Repeated Testing

We have established that one of the most important issues in generalizing results from a sample to the population is the effect of sampling variability. One noteworthy way in which sampling error can mislead an investigator is by the construction (through the random aggregation of subjects in a sample) of a result that can appear to be clinically relevant but at the same time is not representative of the population. The two types of sampling errors that clinical researchers measure are the type I or alpha error rate and the type II or beta error rate, as discussed in Chapter One. The medical and regulatory communities are comfortable with drawing conclusions from concordantly executed studies† when these rates are kept at acceptably low levels.

However, these rates can grow to unacceptably high levels when statistical testing is carried out repeatedly in the same research effort. The monitoring of a clinical study during the course of its follow-up period is a clear example of this phenomenon. Other illustrations are multiple endpoint evaluation, subgroup analyzes, and the evaluation of different contrasts between the arms of a clinical trial with more than two treatment groups. Difficulties with these analyzes have been well elucidated in the literature [2,3,4]. The particular problems induced in the interim monitoring setting have also been elaborated [5].

The principal difficulty with multiple analyzes is that the overall false positive error rate or alpha error rate increases with the number of tests that are executed. Thus, although each test provides the same level of protection, the integrity of the overall process degrades. This is easily demonstrated. Consider an example of a clinical study that assesses the ability of a therapy to reduce the fatal stroke rate. At the conclusion of the research, the investigator plans to construct a test statistic that will produce a type I error rate of 0.05. However, the investigator intends to have the study monitored every year until the five-year study has concluded. At each monitoring point, he is looking for an early demonstration of the same effect that he hopes will be demonstrated at the conclusion of the five-year study. Therefore, a treatment effect finding resulting in a p-value ≤ 0.05 for any of the five analyzes would be sufficient for him.

At first appearance, this collection of tests may appear to offer substantial protection against the occurrence of an alpha error or false positive results. After all, this 0.05 level of protection was satisfactory for drawing conclusions at the end of the research effort. If it will be adequate when applied at the end of the study, why wouldn't it afford adequate protection during the interim monitoring times?

---

* T-tests, chi-square tests, tests of equality of proportions, life table analyzes, and Bayes procedures are but a few of the many types of test statistics brought to bear in the evaluation of clinical research data.

† A concordantly executed study is one that is follows its prospectively written protocol.

The problem is produced by the change in the fundamental event of interest to the investigator. When there was only one evaluation at the conclusion of the study, the focus of his attention was the occurrence of a type I error at the end of the study. With the institution of multiple monitoring, the question is now, "What is the probability that the study produces a false positive result either at the end of the study or during the monitoring procedures?" Because the study could be stopped at every monitoring point, we now need the probability that it meets the criteria for early termination at any one of these points.

Assuming for the moment that these tests (designed to determine the superiority of treatment) are independent of one another,* this probability is easy to identify. Let ξ be the probability that at least one type I error occurs. Then, when there is only one hypothesis test at the end of the study, $\xi = 0.05$. However, when there are five such tests, we must find the probability of at least one type I error. We begin by finding the probability of no type I errors. The **P**[no type I error on the first test] = 1 – 0.05 = 0.95. Thus

$$\mathbf{P}[\text{no type I error on all five tests}] = (0.95)^5$$
$$\mathbf{P}[\text{at least one type I error on the five tests}] = 1 - (0.95)^5 = 0.226.$$

The likelihood of a type I error has increased from 0.05 to 0.226, representing more than a fourfold increase. A more complete examination reveals the rapid increase in the overall type I error rate (sometimes called the familywise error rate or FWER [6, 7]) as a function of the number of tests (Figure 4.2.)Armitage et al. [8] took this one step further by computing the probability that a test statistic that follows the standard normal distribution will exceed any threshold under the null hypothesis. This demonstration revealed that, if an investigator evaluated his data five times during the course of a study, the actual type I error level is 2.5 to 4 times as large as he had planned.† These evaluations are the quantitative justification for the use of the pejorative term "testing to significance".

The underlying reason driving this phenomenon is the random sample. As the sample grows, the data randomly aggregates and disaggregates in different patterns, sometimes making sense, other times not, but always random. As the data set increases, the inclusion of additional data provides continued opportunity for the data points to randomly arrange themselves into the pattern for which the investigator was looking. Just as searching the clouds long enough will eventually reveal the ephemeral shape of an elephant, or a ship moving backwards, the continued interrogation of data, itself randomly aggregated, will reveal the interesting pattern that the investigator anticipated. As Miles pointed out, "If you torture a dataset long enough, it will tell you want you want" [9].

---

* We will see shortly that they are not independent at all, but this scenario is provided solely for illustration.

† This is a basis for requiring smaller type I errors during the interim monitoring procedure.

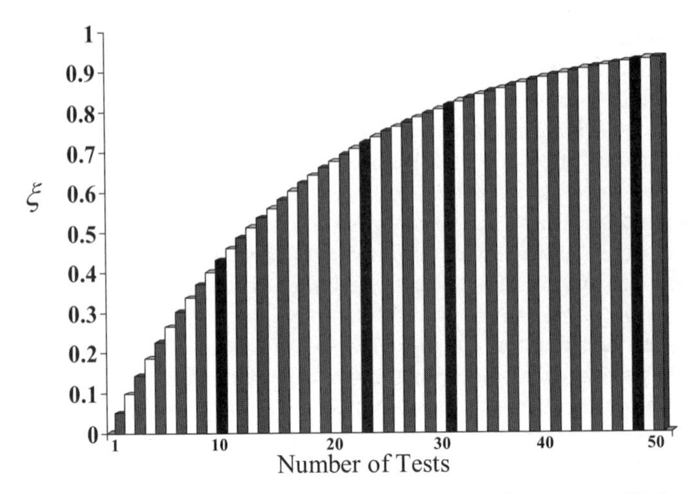

Figure 4.2. Increase in overall type I error as a function of the number of independent tests performed.

4.3.2 Variability

Another phenomenon that, like repeated testing, is produced by the sampling process is the notion of variability. One of the implications of sample-to-sample variability is the following: because one sample's data is different from that of another, then the estimates of quantities produced from one sample, such as a relative risk, will be different than those produced by another sample. Because samples randomly chosen are equally valid, the investigator has no reason to prefer one value of the estimate over that of another. Investigators therefore use this concept of variability to inform them about the accuracy of the estimate.

To statisticians, variability is captured in a measure called the *variance*. This is the degree to which an estimate varies from its long-term average.* To investigators, a useful assessment of variability is conveyed by the 95% confidence interval, a concept that we discussed earlier.†

Monitoring procedures commonly involve a sample's early assessment of a population measure, (e.g., an effect size). An important problem that is induced by the early evaluation of an estimate produced by an incoming data stream is that

* If T is an estimate of a parameter θ, and we denote the long-term average of θ as $E[\theta]$, then the variance of θ is $E[\ T - E\{\theta\}\ ]^2$.The exponent of two is used because it (1)keeps positive deviations of T from its long run average from being canceled out by negative deviations, (2) it is easy to work with mathematically (i.e., the ability to take a derivative is unimpaired), and (3) it is relatively easy to calculate.

† Discussed in Chapter One, Section 1.7.1.

the variability of the estimate is much greater early in a study than it is later in the study.

As an example, consider a clinical trial that is designed to assess the effect of therapy on the intracranial hemorrhage (ICH) rate in adults. Let I_C be the cumulative ICH rate in the control group, and I_A be the corresponding quantity in the active group. It is believed that $I_C = 0.12$, and $I_A = 0.09$, generating a relative risk of $R = 0.09/0.12 = 0.75$. This translates into a $1 - 0.75 = 0.25$ or 25% reduction in the ICH rate that would be attributable to therapy.

However, this is the effect of the therapy in the population at large. From this population, the investigators must choose a series of samples from which they accumulate an estimate of the relative risk. Using Monte Carlo simulation, we can approximate the experience of the investigators in their efforts to accurately estimate the relative risk of therapy in the population (Figure 4.3)

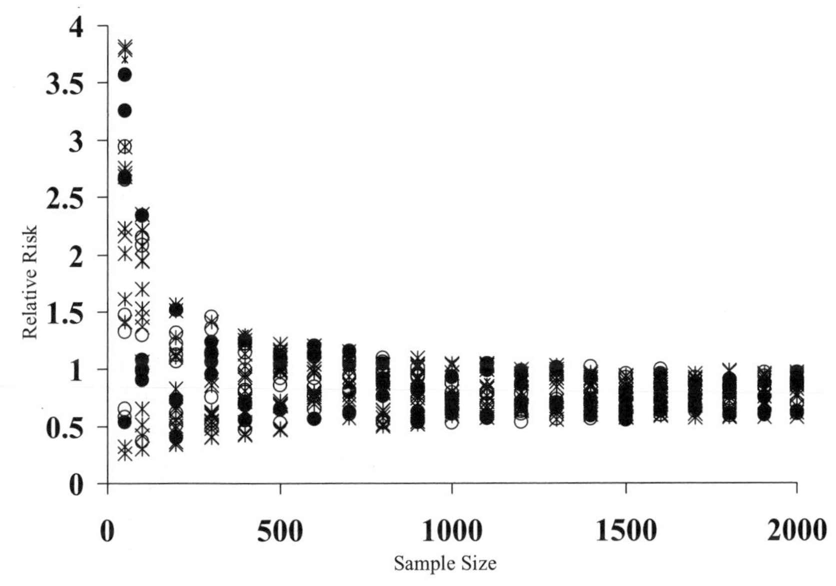

Figure 4.3. Monte Carlo simulation of relative risk estimate as a function of the sample size of the trial. Small sample sizes induce large variability in the relative risk estimate.

From Figure 4.3 we see that drawing samples of observations from a population with a relative risk of 0.75 produces estimates of relative risk that exhibit substantial variability. This variability is most pronounced when the sample size of the study is small. For example, with 100 patients enrolled in the study (50 patients in each of the active and control groups) we observe that the actual estimates of this relative risk range from less than 0.50 to almost 4.0. Thus, even though the relative risk in the population is 0.75, suggesting a moderate benefit, it is quite likely that a population in which the true relative risk is 0.75 could have produced an estimate as large as 3.75. Therefore, an investigator who observes a relative risk in their small sample of 0.75, must question how close this estimate is to the true population relative risk. Although the estimate of 0.75 is on target and within the investigator's expectations, it is not very informative because the population relative risk could be quite different.

As the study progresses, and the sample size increases, the available information from which the relative risk is contracted improves. For example, when the study has recruited and followed 1000 patients (ten times as many patients as were in the study when the range of relative risks extended from 0.25 to almost 4.0), the range of relative risks observed becomes much smaller (0.50 to 1.25). This is a much more precise estimate of the relative risk.

The fact that a larger sample size produces a more precise relative risk is intuitive, and the implications for the simple application of effect size threshold for identifying a relative risk are clear. Important effect sizes when identified very early in a study, are remarkably unreliable and cannot be embraced as definitive. Thus, monitoring rules must reflect this observation by demanding a much larger effect size earlier in the study than would be required later when the information is more precise and reliable. This is reflected in the shape of the curves of Figure 1.2 in Chapter One, reproduced here (Figure 4.4.

he lines in Figure 4.4reflect the magnitude of the test statistics that is required for recommending that the study be terminated. It reveals that a greater strength of evidence is required in the early part of a study to suggest early termination. This requirement is in part due to the observation that treatment effect measures are much less reliable early in a study.

4.3.3 Dependence

The application of repeated hypothesis testing on an incoming stream of data from an ongoing clinical research program produces two problems that we have discussed thus far. The first is the inflation of the sampling error rates that is induced by the repeated testing. The second is the unreliability of estimates that are based on very small samples. A third issue which requires special consideration is that commonly, the different examinations of the data are dependent on one another.

This is the concept of dependency that we saw in Chapter Two. There, this idea of dependence was demonstrated through the evaluation of conditional probability, a tool that will be useful for us here as we begin the development of basic interim monitoring procedures. Dependence arises in the interim monitoring situation because the interim evaluation of clinical research involves the sequential examination of accumulating data. Data that is evaluated at a current monitoring point contains new data that has not yet been assessed plus data that has been evaluated in an earlier monitoring appraisal. The inclusion of this data that has already been assessed induces a dependence between the two evaluations.

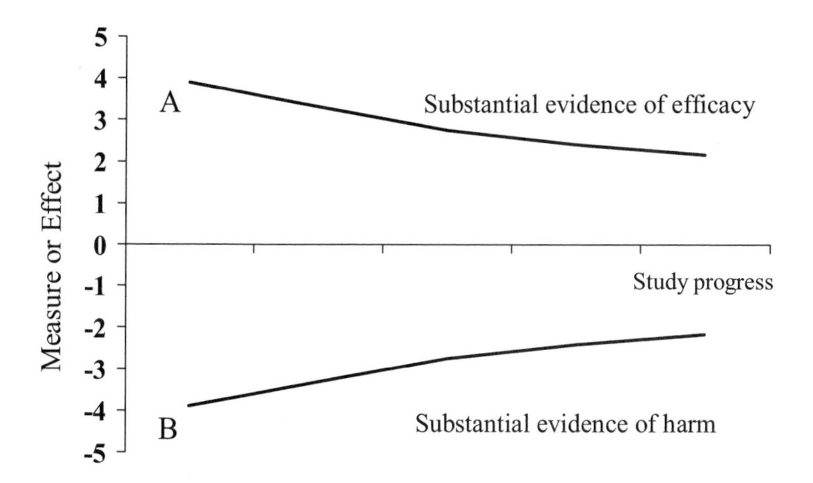

Figure 4.4. Example of monitoring boundaries for efficacy and harm taken from Chapter One (Figure 1.2). Crossing the upper boundary (A) suggests that there is substantial evidence of efficacy, whereas crossing the lower one (B) suggests substantial evidence for harm.

A simple way to appreciate this dependency concept is to consider the following hypothetical circumstance. An investigator plans to end a study when she has concluded an evaluation of 50 patients. The investigator takes an "interim look" at her study results when 49 patients have completed their evaluation. Knowledge of the result when n = 49 provides useful information for her when n = 50. In fact the assessment when n = 50 has changed very little from the evaluation when n = 49, and our intuition tells us that the "early" evaluation has all but completely informed her about the results at the conclusion of the study. The early examination does not tell her about the result of the 50th patient, but it does inform her about the cumulative study finding of all 50 patients.

Acknowledging the presence of dependence begs the question of how this dependence can be used to the investigator's advantage. We can now demonstrate how this dependency can be used to formally "update" a probability of interest for us. Consider an investigator who is interested in measuring left ventricular ejection fractions (LVEF) for a series of patients who have diabetes mellitus. The patients she observes are from a population that the literature suggests has a mean LVEF of 65 and a standard deviation of 7. However, the observer notes that many patients have LVEFs greater than 65; she would like to understand the reason for the discrepancy between her observation and the literature. She knows, for example, that, through the influence of sampling error, a population with a mean LVEF of 65 can produce a sample with a larger mean. However, she also understands that the greater the difference between the sample mean and 65, the less likely the explanation that the large sample mean was due to the play of chance; it is more likely that the population mean itself is greater than 65.

During the design phase of this study, the investigator computes some probabilities of interest. She assumes that her data follow a normal distribution with mean 65 and standard deviation of 7. Call $\overline{X}_m$ the sample mean LVEF based on m observations. Then, at the end of the study $\overline{X}_{100}$ follows a normal distribution with the same mean $\mu = 65$. The variance of the mean is $\sigma^2 = {}^{49}\!/\!_{100} = 0.49$. With this as background, she can make statements about the likely values of $\overline{X}_{100}$ based on this distribution. For example, she can compute the probability that $\overline{X}_{100} \geq 67$ if the underlying population mean is 65 as

$$\boldsymbol{P}\left[\overline{X}_{100} > 67\right] = \boldsymbol{P}\left[\frac{\overline{X}_{100} - 65}{\sqrt{0.49}} > \frac{67 - 65}{\sqrt{0.49}}\right] = \boldsymbol{P}\left[N(0,1) > 2.86\right] = 0.002.$$

Thus, it is unlikely that the mean LVEF of 100 patients will be greater than 67 if the population mean is 65. The observation of a sample mean greater than 67 is therefore, evidence that, in a concordantly conducted research program, the population mean is likely to be greater than 65.

The investigator plans to recruit 100 patients for the study, but will monitor the results of the study (in this case, examine the sample mean) when she has 50 patients enrolled. Assume that she observes $\overline{X}_{50} = 70$. How likely is it that at the conclusion of the study, the investigator will observe $\overline{X}_{100} \geq 67$ given $\overline{X}_{50} = 70$? From Appendix A, we see that $\overline{X}_n$ still follows a normal distribution but with mean μ_c and variance v_c, where

$$\mu_c = \mu + \frac{m}{n}\left(\overline{X}_m - \mu\right)$$

$$v_c = \frac{n-m}{n^2}\sigma^2,$$

where $0 < m \leq n \leq p$. Thus, the new mean for $\overline{X}_n$ that is now conditioned on the value of $\overline{X}_m$ depends on both m and $\overline{X}_m$. The variance of $\overline{X}_n$ that is conditioned on the value of $\overline{X}_m$ depends on m.

Assume that the investigator knows that the mean of the first 50 observations is 70, that is, $\overline{X}_{50} = 70$. The mean and variance of $\overline{X}_{100}$ given $\overline{X}_{50} = 70$ are

$$\mu_c = 65 + \frac{50}{100}(70-65) = 67.50.$$

$$v_c = \frac{100-50}{(100)^2}49 = 0.245.$$

She can now compute

$$\boldsymbol{P}\left[\overline{X}_{100} > 67 \mid \overline{X}_{50} = 70\right]$$

$$= \boldsymbol{P}\left[\frac{\overline{X}_{100} - 67.5}{\sqrt{0.245}} > \frac{67-67.5}{\sqrt{0.245}}\right] = \boldsymbol{P}\left[N(0,1) > -1.01\right] = 0.844.$$

Note that knowledge of the prior mean left ventricular ejection fraction has changed the probability that the mean at the end of the study will be greater than 67. Without knowledge of the value of $\overline{X}_{50}$, the probability was 0.002. With this information, the probability is 422 times greater. The important new information about the first 50 observations has changed the anticipated mean of the 100 observations (Figure 4.5)

The preceding computation begs the question of how large would the sample mean based on 50 observations have to be in order to have at least a 95% chance of having the mean based on 100 observations be greater than 67. This is a useful question that gets to the heart of interim monitoring. By identifying such a boundary for $\overline{X}_m$, the investigator can consider terminating the study if this boundary is reached, because it is very likely that, had she continued the study, $\overline{X}_{100}$ would be greater than 67. The general result is provided in Appendix A, from which we see that

$$b = \frac{n}{m}\left[67 + 1.645\sqrt{\frac{n-m}{n^2}\sigma^2} - \mu\right] + \mu.$$

In this case $n = 100$, $m = 50$, and $\mu = 65$. We can compute

$$b = \frac{100}{50}\left[67 + 1.645\sqrt{\frac{100-50}{100^2}49} - 65\right] + 65 = 70.63.$$

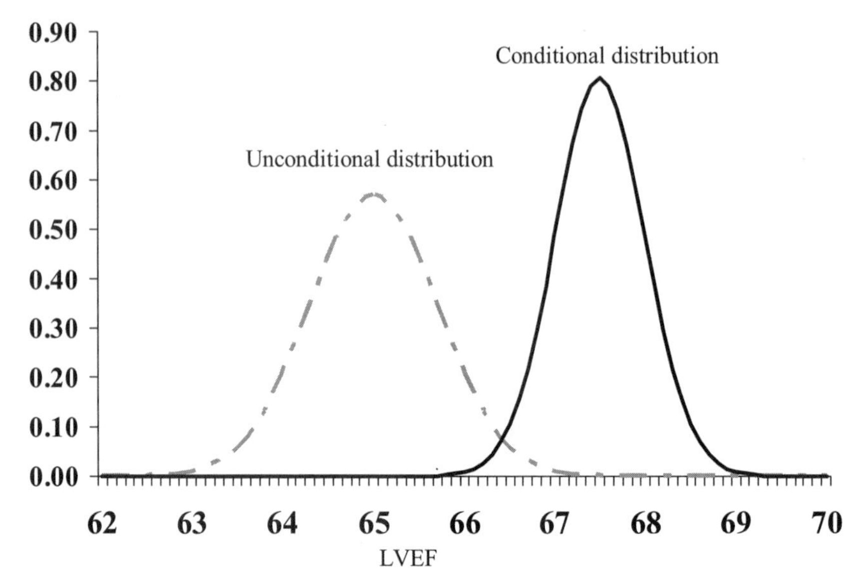

Figure 4.5. The unconditional distribution of the mean LVEF from 100 observations has a mean = 65 and a variance = 0.49. The conditional distribution based on knowledge of the first 50 observations have a larger mean of 67.5, and a smaller variance of 0.245.

Thus, if the mean based on 50 observations is at least 70.63, the probability that the mean at the end of the study, $\overline{X}_{100}$, is greater than 67 is 0.95. If this probability is deemed high enough, then the investigator might consider stopping the study when 50 subjects are enrolled and $\overline{X}_{50} \geq 70.63$.

The identification of this level of dependence has allowed us to link a past event (in this case, $\overline{X}_{50} \geq 70.63$) with a future event ($\overline{X}_{100} > 67$) and therefore, compute a threshold for action based on a value of a statistic at a monitoring point. This is a type of computation that we will come back to in succeeding chapters.

4.4 Following Trajectories

The fundamentally new statistical principle for us in monitoring clinical research is the concept of path analysis. It is perhaps intuitively obvious that tracking the magnitude of the treatment effect over time would be a useful feature in the ongoing assessment of clinical research. However, the interpretation of these trajectories can be complicated. We already know that we must keep in mind the issues of multiple testing and variability, although simultaneously incorporating the notion of dependency considerations.

We will now identify some additional characteristics of these treatment effect trajectories, allowing us, in the end, to apply some of the discussion of Chapter One.

4.4.1 Riding the Path of an Effect Size

We can begin our discussion of trajectories in clinical research by considering a hypothetical clinical trial that is designed to assess the effect of a therapy that the investigator believes will reduce the incidence of fatal strokes in patients who have diabetes mellitus. The investigator plans to randomize patients to either active or control group therapy and then follow these patients for a year. At the conclusion of the research, he plans to compare the cumulative fatal stroke rate in the active group with that of the control group.

However, suppose that we have a unique vantage point. Rather then being required to wait until the end of the study to observe the difference in the number of fatal strokes between the control group and the active group, we are instead permitted to watch this difference develop on a daily basis for the duration of the study. If we were permitted to "ride the effect size path", tracking its every movement, what would it look like?

To help in this depiction, define $S_A(t)$ as the number of fatal strokes that occurs in the active group at time t, and $S_C(t)$ the number of fatal strokes that occurs in the control group at that time. We will use a very elementary definition of effect size, $S_C(t) - S_A(t)$, tracking this difference as time passes. At the beginning of the trial $t = 0$, and $S_C(0) = S_A(0) = 0$. As recruitment begins, patients receive their treatment and start to accrue follow-up time. Fatal strokes begin to occur, and the difference $S_C(t) - S_A(t)$ immediately registers them. If the number of fatal strokes in the control group is greater than in the active group, our difference becomes positive. Alternatively, the difference becomes negative when there are more fatal strokes in the active group than there are in the control group.

However, clinical events are not very predictable; in fact, they occur randomly during the follow-up period, and will commonly aggregate just through the play of chance (Figure 4.6).

Figure 4.6 demonstrates that the path followed by the difference $S_C(t) - S_A(t)$ wanders. If there has been a recent sequence of fatal strokes in the control group, the difference will inch its way upward. If a recent cluster of fatal strokes occurred among patients in the active group, then $S_C(t) - S_A(t)$ will decrease, perhaps becoming negative. With no consistently positive or negative force, this difference will meander in one direction then in another.

There are many possible trajectories that $S_C(t) - S_A(t)$ could follow. After its initial value of 0, it could take off rapidly in a positive direction, or actually do the reverse and become rapidly negative as the number of fatal strokes in the active group exceeds the number of these events in the control group. Alternatively, the difference $S_C(t) - S_A(t)$ could hover around zero. However, the number of available paths that this difference $S_C(t) - S_A(t)$ could follow is too large to tabulate. With each new day, the location of $S_C(t) - S_A(t)$ could be different, spawning a new collection of paths that are too numerous to begin to count.

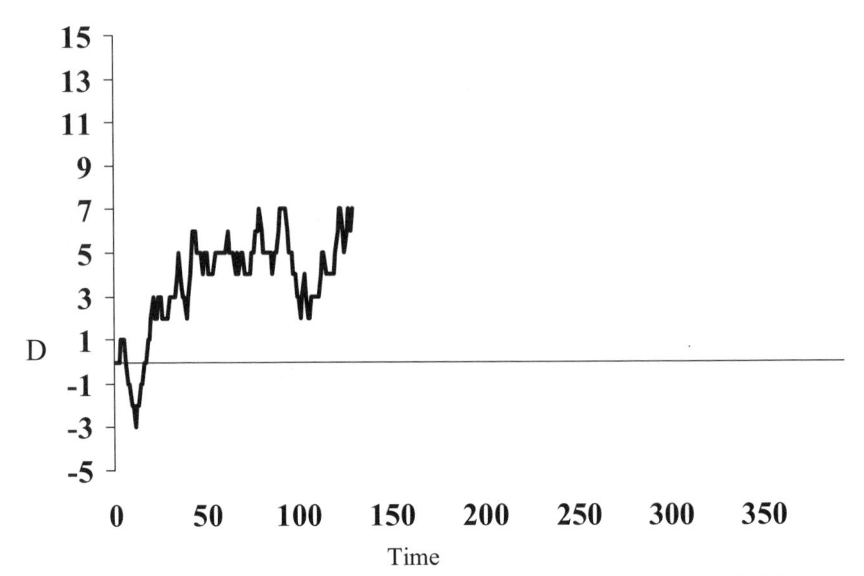

Figure 4.6. Cumulative difference in the number of fatal strokes in a clinical trial from day 0 to day 129. $D = S_c(t) - S_t(t)$

This is a phenomenon that we saw in Chapter Two. Thus, if we wanted to consider assigning a probability to each of these paths, we would find that just as we could not assign positive probability to each point in an interval on a real line, we cannot assign positive probability to each possible path of $S_C(t) - S_A(t)$.

From Figure 4.6 we see that as time progresses, the number of fatal strokes in the control group is exceeding the number of active group fatal strokes. From our vantage point, we cannot help but begin to feel some excitement. After a dip in the difference that led to $S_C(t) - S_A(t) = -3$ early in the study, the number of fatal strokes in the control group has consistently exceeded those of the active group, producing a positive value of $S_C(t) - S_A(t)$. If this difference is sustained (and from Figure 4.7 this is precisely what appears to be happening), the feeling that the trial is demonstrating a beneficial effect of therapy begins to develop momentum itself. Natural questions, for example, "Has this difference been sustained long enough, and is there an extreme point (or extreme trajectory) that this difference could follow that would suggest that we "do something" about the study?" arise. However, when allowed to continue, we observe that the trajectory of $S_C(t) - S_A(t)$ abruptly changes (Figure 4.7).

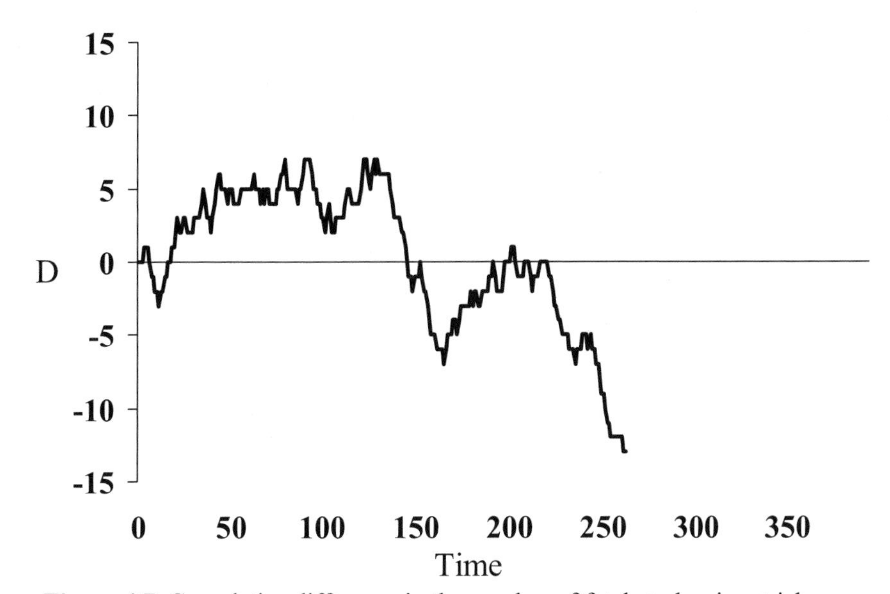

Figure 4.7. Cumulative difference in the number of fatal strokes in a trial from day 0 to day 262 $D = S_c(t) - S_t(t)$

The positive values of $S_C(t) - S_A(t)$ which have produced and sustained a beneficial trajectory have now been followed by a relative increase in the number of deaths that have been sustained in the active group; this new tendency produced a decrease in the value of $S_C(t) - S_A(t)$. However, the underlying assumption that produced this graph is that there is no difference in the population between $S_C(t)$ and $S_A(t)$. The fluctuations in $S_C(t) - S_A(t)$ are those induced by chance alone, and not by any change in the execution of the study. There was no alteration of the entry criteria of the study, nor was there a change in the follow-up procedures mandated by the protocol. The endpoint committee whose task it was to determine whether a death was due to a fatal stroke has not been unblinded to therapy assignment, and no other aspect of that committee's modus operandi has altered. As far as we can tell, there is no reason for the trajectory alteration.

Early optimism is now replaced by concern as the number of fatal strokes in the active group exceeds those in the control group, and pressure again builds to once again "do something" about the study. However, continued observation produces yet a different trajectory (Figure 4.8).

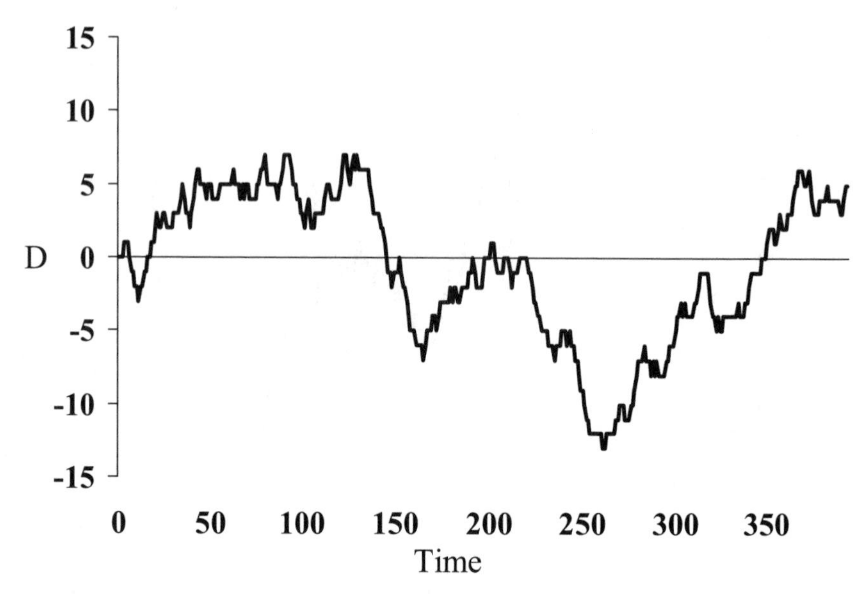

Figure 4.8. Cumulative difference in the number of strokes in a clinical trial over the entire course of the study. $D = S_c(t) - S_t(t)$

This figure reveals yet another important and radical change in the direction of the difference in fatal strokes $S_C(t) - S_A(t)$. A final surge in the number of control group fatal strokes has increased the difference in the number of these events between the two groups, leading to a difference that is close to zero at the end of the study.

4.4.2 The Direction of Random Drift

If there were no therapy effect on the fatal stroke rate, we might expect that the difference $S_C(t) - S_A(t)$ would hover around zero, moving slightly positive, then "adjusting" to turn slightly negative over time. However, this is not the case. The "adjustments" commonly occur only after large excursions. Typically, large positive excursions are followed by large negative excursions.

Thus, in the absence of a treatment effect, there will nevertheless be marked, even extreme excursions in $S_C(t) - S_A(t)$ over time. This property can be cause for concern, because this random movement will "mimic" the appearance of a treatment effect. However, in the absence of a treatment effect, if enough time is permitted to elapse, large positive excursions will be matched by large negative ones. This will lead to an "average" value of the treatment effect that is close to zero. However, this small average value masks the appearance of trajectories that contains wide negative ones (Figure 4.9. *

* This is not an unfamiliar setting. For example, the mean temperature of Indianapolis, Indiana is 47 degrees Fahrenheit. This moderate value does not reveal that the temperatures in

Figure 4.9. Four trajectories of the estimate of efficacy in the absence of a treatment effect.

The last panel of Figure 4.9 is particularly illustrative of the need to follow the process for sufficient time. An investigator might assume from this panel that the effect of therapy was harmful out to 350 days. However, the data in the panel was generated assuming no difference in the effect of the therapy on stroke. If the process were observed "long enough" the negative trend will reverse.

4.4.3 Two Path Influences

This demonstration reveals several interesting properties of effect size trajectories over the course of the study. There are two influences that would affect the path or trajectory of $S_C(t) - S_A(t)$ over time. The first is the treatment effect. If the active therapy is very effective, we might expect an early, rapid, and sustained decline in the path of $S_C(t) - S_A(t)$ over time. The random movement induced by the play of chance remains in effect; however, if the treatment effect is large, $S_C(t) - S_A(t)$ will tend to get larger over the course of time. Similarly, a harmful effect of the active treatment might produce an early increase in the difference.

The second influence on the path of $S_C(t) - S_A(t)$ is a random one. This random influence is the explanation for the roughness or bumpiness of the trajectory. The investigator may have confidence in his ability to predict the fatal stroke rate at the conclusion of the study and therefore may have very good and useful

---

the winter can plunge to 25 degrees below zero, and that summertime temperatures commonly exceed 90 degrees.

estimates of the difference in the number of fatal strokes between the two groups at the study's conclusion. However, he most certainly cannot predict what new strokes that he will learn about on any given day. This is influenced by a host of factors: some known, but many unknown. The ensemble of these influences is a random, unpredictable effect on the difference $S_C(t) - S_A(t)$. Thus, the effect of therapy influences the long-term trajectory of $S_C(t) - S_A(t)$ and the daily movement is visibly affected by the random influences.

4.4.4 Trajectory Concerns

An issue that is of concern is the large swing in the path of the difference in the fatal stroke rate between the control and active group as the trial progressed. Positive trajectories of the $S_C(t) - S_A(t)$ are followed by negative ones. In this study, in which there is no effect of the intervention on the fatal stroke rate, we anticipated that the difference in strokes between the two groups $S_C(t) - S_A(t)$ would be close to zero. However, this was only true in the coarsest sense of the term. The difference in the end wound up near zero. However, the path was punctuated by excursions away from zero. Another example reveals a trajectory over three years of time produced from an evaluation in which there is no effect of therapy (Figure 4.10).

Again, large positive movements are followed by large negative movements. These swings or excursions become greater and greater as the trial proceeds. Thus, although it is true that $S_C(t) - S_A(t)$ remains small when averaged over the duration of the trial, it by no means stays close to zero for the duration of the study. This is a cause for concern. One can hardly consider terminating a study early based on the appearance of a large treatment effect, if these large effects are likely to be mimicked in studies where the effect of therapy is absent.

4.4.5 Memoryless

If anything comes as a surprise in these demonstrations, it is the appearance of these wild deviations in the value of $S_C(t) - S_A(t)$. Under the assumption of no treatment effect, we would expect the value of this difference to be close to zero in the long run, and it is only natural to assume that $S_C(t) - S_A(t)$ will tend to "stay close" to zero throughout the duration of the study.

The preconception that the values of $S_C(t) - S_A(t)$ will stay near zero comes from the recognition that there is no systematic influence pushing it away from zero. In the absence of this pressure, we anticipate that the difference will remain close to zero (i.e., it has inertia). However, just as there is no consistent force that pushes it away from zero, neither is there a stabilizing power that keeps the difference near zero. In the absence of these consistent and sustaining influences, the difference $S_C(t) - S_A(t)$ fluctuates wildly. It meanders well away from zero, only to eventually return to zero, after which it moves away once again. These counterintuitive excursions are the hallmark of a process that is momentumless. We say that the process is *memoryless*.

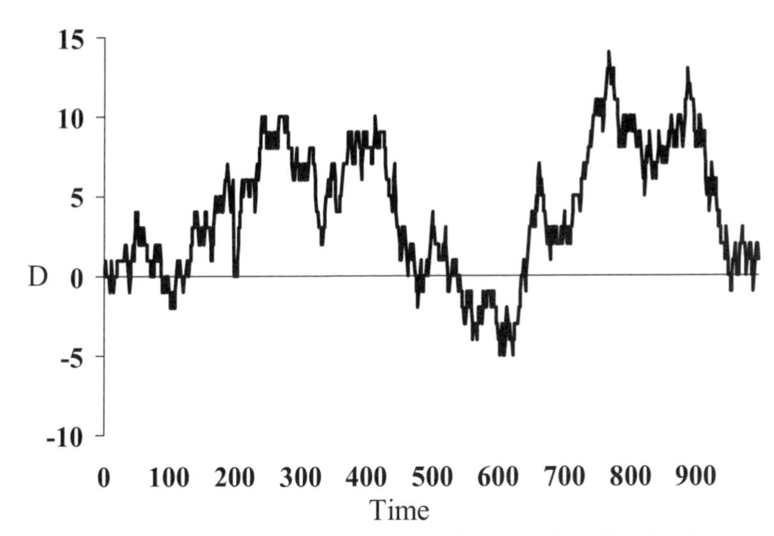

Figure 4.10. Projected cumulative difference in the number of strokes in a clinical trial over the entire course of the study lasting over 900 days. $D = S_c(t) - S_t(t)$

A process is memoryless if an examination of its prior trajectories does not inform us about its future. Memoryless processes can make us feel somewhat uncomfortable because we are used to being able to examine the history of a system in order to gain important information about its future behavior. In fact we create systems that contain and embed useful memories, thereby permitting us to learn about the future based on an examination of the past.

As an example, consider the relationship that develops between spouses. Their history is a long and intertwined one. Over the years, they develop joint memories and experiences that are rich with understandings, misunderstandings, apologies, and illumination. They learn when it is best to speak up and when it is best to stay silent. Over time, each partner learns about the other, observing, understanding, and then predicting the reaction of each other to various circumstances. This is a process that is rich with informative history and deep memories, allowing one member of the relationship to react to and thereby influence the other.

A memoryless alternative would be the situation in which one wakes up every morning to face a new different partner with whom one has no past experience. With no useful memories to guide you, how do you proceed in your interactions with the new partner? You have no history with this person, no shared experiences, no understanding. What do you know about him? How can you relate to her? How do you make decisions together when the reaction to anything that you do or say is unpredictable? You develop no useful momentum with the person because anything that you may have happened to learn about him or her that day is undone by the presence of a new partner the next morning when you start the process all over again. This is a memoryless process. The product of memoryless processes is wide, unpredictable swings.

Monitoring rules in clinical research are commonly based on memoryless processes and we have to develop new intuition to work satisfactorily with them.

4.4.6 Alternative Measures

The previous sections demonstrated the wild swings that can occur in the path of a treatment effect over time. These swings that we now know typify a memoryless process can complicate any attempt to draw useful and reliable conclusions from the observations of extreme trajectories. A natural alternative to monitoring the treatment itself would be to monitor the test statistic. In the example of the hypothetical clinical trial that is designed to monitor stroke rates, this would mean that we need to monitor not $S_C(t) - S_A(t)$ but

$$\frac{S_C(t) - S_A(t)}{\sqrt{Var\left(S_C(t) - S_A(t)\right)}}.$$

The difference between the path that is followed by the treatment effect $S_C(t) - S_A(t)$ and that of the test statistic is profound (Figure 4.11). We see that, although the path of the treatment effect is subject to large fluctuations, the test statistic's trajectory is substantially moderated. It is by no means a smooth path, but division by the standard deviation of the difference buffers and protects its trajectory from large excursions.

However, although the use of the test statistic appears to confer an advantage to the monitoring process, we must observe that we can be misled by this moderation as well. Recall that, earlier in this chapter, we discussed the variability occurring in samples based on small amounts of data. Because populations in which there is no treatment effect can produce small samples that have large treatment effects, we cannot rely on the findings of a treatment effect in a small sample, and must wait until the sample is larger and a more precise estimate is produced.

Examining the trajectory of the test statistic does not convey this important principle. In fact, the test statistic, by being a "normed" treatment effect, is specifically designed to remove the effect of variability. It does this by dividing the treatment effect by its standard deviation at each point in time so that its variance is one. This division occurs early in the trial when the variance is large, as well as later in the trial when the variance is small.

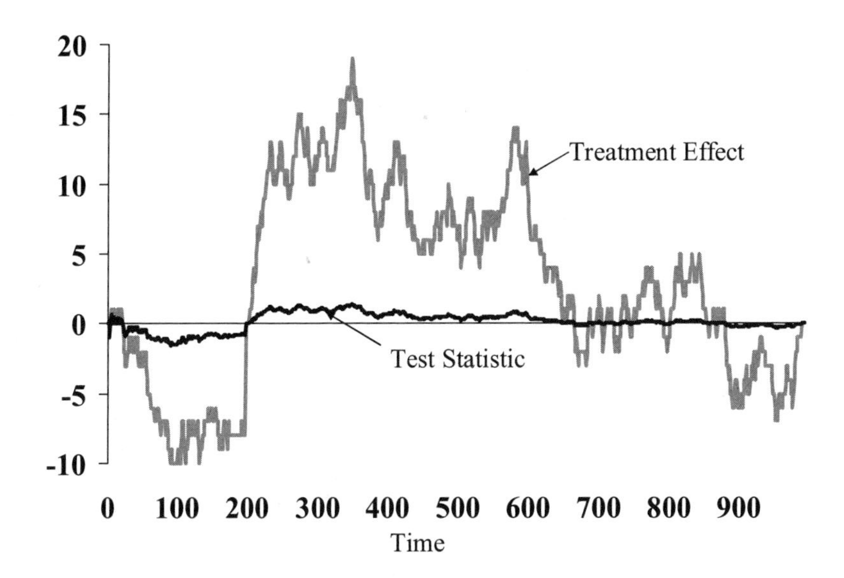

Figure 4.11. The trajectory of the test statistic over time has a more moderate trajectory than that of the treatment effect.

Thus, by scaling the treatment effect in this matter, its graph conveys the false notion that the precision of the treatment effect is the same, when we know, to the contrary, that when the sample is small, the precision of the treatment effect estimate is unacceptably poor.

4.4.7 Dilemmas and Resolutions

The preceding examinations have provided some useful insight into how the monitoring of a treatment effect might usefully guide the decision to terminate a clinical trial early. However, these same examinations have uncovered some complexities in monitoring treatment effects. We saw that there is substantial variability that surrounds the estimate of the treatment effect. In addition, the process is a memoryless one, producing wide unpredictable swings in the trajectory of the treatment effect. These two influences tend to make the treatment effect's path very difficult to predict. In fact, the large excursions that the treatment effect experiences get even larger over time.

We have also seen that following the trajectory of the test statistic is not the solution because, by incorporating variability, it provides a false assurance of the homogeneity of the precision of the estimate of the treatment effect. Finally, we have the multiple comparisons issue to address, an obligation that we must meet if we are to successfully control sampling error in drawing conclusions.

The previous consideration of the issues raised by following a treatment effect over the course of time raises special needs for the monitoring tool that we can now articulate. Specifically, we require a measure that incorporates (1) the magnitude of the treatment effect itself, (2) the different degree of variability of the treatment effect that occurs over the course of study, (3) the likelihood of important excursions in the trajectory over time, and (4) the memoryless property. Fortunately, we have already been exposed to such a process. Each of these properties that we attribute to the excursions of a treatment effect over time is already contained in the water-suspended movements of a minute pollen grain.

4.5 Brownian Motion

As pointed out in Chapter One, over the course of approximately 100 years, several scientists observed, described and then mathematically derived the properties of Brownian motion. In this section, we will discuss some of the mathematical details that are most useful for the application of this process to the development of monitoring rules in clinical research.

Our focus is on the description of the movement of an *element*. That element may be a speck of pollen, or the trajectory of a function of a treatment effect in a clinical study. Specifically, we will concentrate on identifying the position of that element at a particular time t; we will denote the location of that element $B(t)$. Our discussion is limited to one-dimensional movement, that is, the movement of the element is either up or down, over the course of time. Thus, the expression $B(0) = 0$ states the location of the element at time 0 is 0 (or at the origin). The statement $B(1) = 2$ says that the location of the element at time 1 is 2. The statement $B(2) = -1$ states that the location of the particle at time 2 = –1 (Figure 4.12).

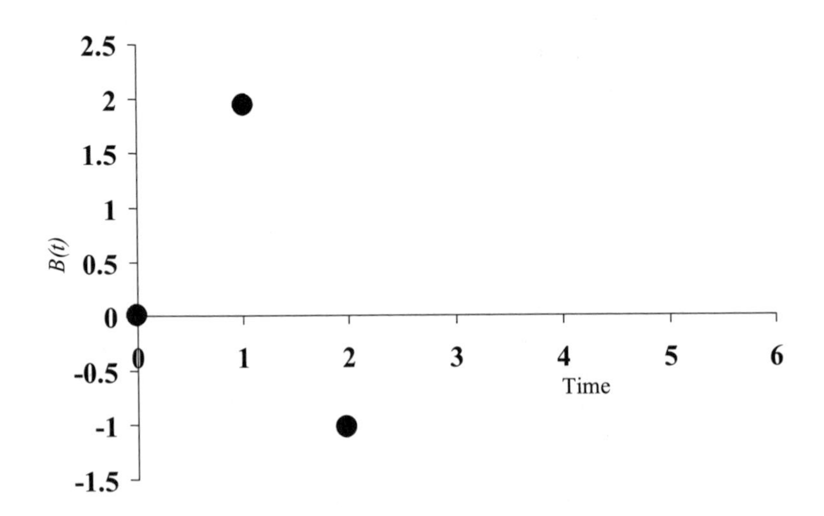

Figure 4.12. The location $B(t)$ of the element at three points in time.

After a description of the mathematical properties of standard Brownian motion, we discuss how one actually computes probabilities for the location of the element that exhibits Brownian movement, and apply these probabilities to relevant computations in clinical research.

4.5.1. Standard Brownian Motion

The mathematical properties of Brownian motion are well defined, and our discussion will be limited to its very elementary, central properties.* However, as we enumerate and begin to work them, we will also be sure to link these properties to clinical research monitoring, the central area of application here. The following elementary discussion of Brownian motion is taken from [10].

4.5.1.1 Property 1: B(0) = 0

Although it is theoretically possible for the process to have any value at any point in time, the movement must begin from some place. Standard Brownian motion begins at location 0 at time $t = 0$. This is consistent with the magnitude of the treatment effect at the beginning of a study.

4.5.1.2 B(t) Is Normally Distributed

The position of the element at some future time t can not be known until that time arrives. However, predictions can be made about the likely location of the element. Specifically, standard Brownian motion follows a normal distribution. The mean of this distribution is 0, and the variance is t. We say that $B(t) \sim N(0, t)$.

The invocation of the normal distribution here is comforting because it is a distribution that is ubiquitous, easy to work with, and has properties with which we are comfortable. However, the variance of $B(t)$ requires special attention. It is not constant (unlike the mean, which is always zero), but actually increases in time. If we are at time 0, we are relatively assured where the element will be at time 0.001. We are less sure of the location of the particle at time 1, and are much less sure of the particle's location at time t = 100. The farther into the future that we wish to predict, the less certain we are of the element's location at that future point (Figure 4.13).

Figure 4.13 provides a sense of the probable positions of the element for time points t_1, t_2, t_3, and t_4 when examined from the starting time $t = 0$. The rotated normal curves provide a sense of the certainty about the element's location. When only a relatively short time has elapsed from $t = 0$ to $t = t_1$ the element has not had much time to move away from its initial location of 0, and the narrow curve reflects our high confidence that the element is not far from 0.

* There are other properties of Brownian motion that, although not of direct relevance to us in these more general discussions nevertheless retain interest. For example, Brownian motion is an incredibly jerky process with the element changing its location instantaneously and randomly. Thus, although line segments can be drawn between adjacent locations of the element, these segments are so short that their slopes cannot be computed. Mathematicians describe this process as being continuous but not differentiable.

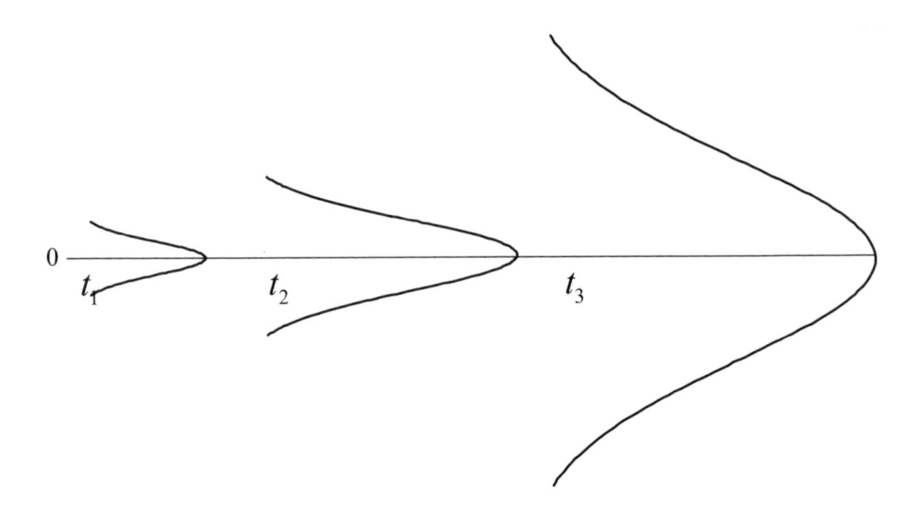

Figure 4.13. Probability distribution of location of a Brownian element. As time increases, the distribution disperses, with the more extreme locations becoming more probable.

By the time $t = t_2$, quite a bit more time has elapsed, and, although the most likely interval of locations for the particle remains around zero, there is now a substantial probability that the element has moved away from zero. This tendency towards a greater likelihood of locations of the element away from zero increases as the elapsed time increases from t_1 through t_2 to t_3.

This property of increasing variance with increasing time is an appropriate model to use for monitoring many clinical effects in a research effort over time. We are more certain of the cumulative magnitude of the treatment effect a short distance into the future because, even though it is impossible to predict the size of the treatment effect contained in the data that has not yet appeared, its impact on the cumulative treatment effect observed thus far will be relatively small. However, the farther out in time we wish to project, the greater the quantity of data yet to arrive, and the greater the impact of this data on the cumulative treatment effect. Thus, we are less and less certain of the magnitude of the treatment effect as we make more distant projections.

4.5.1.3 *B(t)* Is Memoryless

We discussed the memoryless feature of Brownian motion in Section 4.5.4 when we explored the reasons for the reversals in the large excursions of the treatment effect over time. This notion of memoryless is commonly expressed mathematically through what is known as the *independent increment property*. Consider two time intervals $[t_1, t_2]$, and $[t_3, t_4]$ where $t_1 < t_2 \leq t_3 < t_4$. Then the change in the position of the element from time t_1 to t_2, denoted by $B(t_2) - B(t_1)$ provides no information about the change in the location of the element from t_3 to t_4, denoted by $B(t_4) - B(t_3)$. We say that the change in the location of the element $B(t_4) - B(t_3)$ is independent of an earlier change in its position $B(t_2) - B(t_1)$. Because future changes in the element's location are not linked to past changes, the movement of the element is momentumless.

4.5.2 B(t) Conditioned on the Past

Although a movement of a Brownian element in the past provides no information about the future movement of the position, a past position can be very useful in predicting a future position of the particle.

As an example, suppose that we are interested in the position of the element at time $t = 100$. If the current time is $t = 0$, then all that we know about the element's location at time $t = 100$, a point that is far into the future, is that the element's position at that time will follow a normal distribution with a mean of 0 and variance 100. Essentially, the position of the particle could be almost anywhere at time $t = 100$ (Figure 4.14, Panel 1).

However, suppose we are now at time $t = 99$, and we know that B(99) = 7. This information should inform us about the likely position of the element at $t = 100$. After all, the particle has only one time unit to change its location. Although any magnitude of movement is possible, we do not expect that in this short period of time, the particle will radically change its position. We would therefore, expect that the particle would remain close to 7 (Figure 4.14, Panel 2).

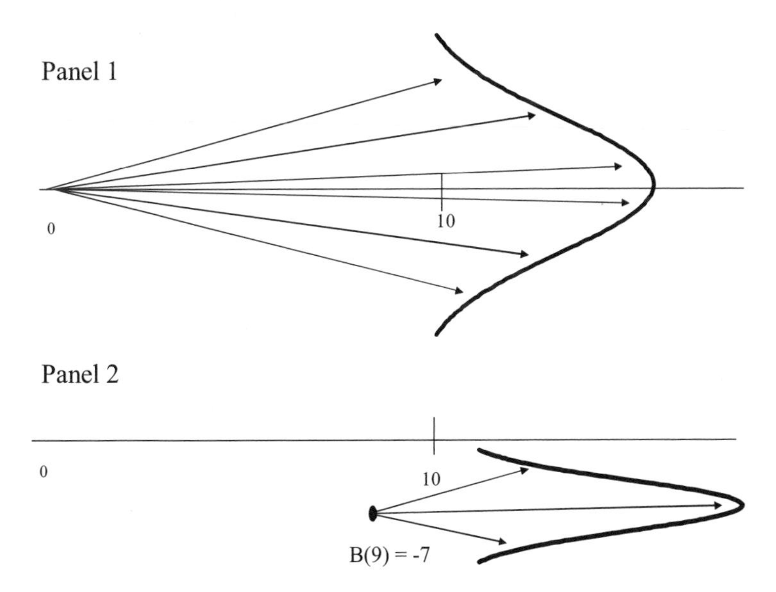

Figure 4.14. The future location of the element is much less certain when attempting to predict ten time units into the future (Panel 1), than when trying to predict one time unit into the future (Panel 2).

It is this type of reasoning that is invoked when predicting the future position of a Brownian element based on information about its past.

This concept of *Brownian motion conditioned on the past* may be stated mathematically as follows. If we are given two time points t_1 and t_2 such that $t_1 < t_2$, and we know the location at the first time point, that is, $B(t_1) = a$, then the probability distribution of $B(t_2)$, although remaining normally distributed, has different parameters. Specifically, its mean is no longer zero, but is now a and its variance is $(t_2 - t_1)$. This is commonly stated as the translation property of Brownian motion. It is as though the Brownian element, having arrived at $B(t_1) = a$, starts again, but now all future positions must be relative to a, and all future times must be based on a starting time of not 0 but t_1. Applying this to the previous example, we compute that the probability distribution of the location B(100) when we know that B(99) = –7 is a normal distribution with mean –7 and variance 100 – 99 = 1. This provides a much more precise estimate of the element's location than what could be obtained from the starting point of the process at $t = 0$.

As another example, suppose that we wish to find the probability that a Brownian element's future location is greater than b at time t_2 (i.e., $B(t_2) \geq b$) given that we know the current location at time t_1 is $B(t_1) = a$. We simply use the normal distribution to compute

$$\begin{aligned} \boldsymbol{P}\left[B(t_2) \geq b \mid B(t_1) \geq a\right] &= \boldsymbol{P}\left[N(a, t_2 - t_1) \geq b\right] \\ &= \boldsymbol{P}\left[N(0,1) \geq \frac{b-a}{\sqrt{t_2 - t_1}}\right] \\ &= 1 - F_Z\left[\frac{b-a}{\sqrt{t_2 - t_1}}\right], \end{aligned}$$

where $F_Z(z)$ is simply the probability that a standard normal random variable is less than some value z.

When first considering this concept of using past information about Brownian motion to predict the future, it may seem that we have arrived at a contradiction. We have said that Brownian motion is memoryless, yet it would seem that the past "memory" of the element's location at time t_1 has helped us to predict its future location at time $t_2 > t_1$. However, the "memoryless" in the memoryless process refers to memory about momentum. A rapid increase that the particle experienced in the past does not affect the likelihood of a rapid increase in that element's position in the future. The most recent past position is valuable in making future predictions. How the element reached that position (either by an increase or by a decrease) is irrelevant.

Another way to say this is, if we have (1) a sequence of time points t_1, t_2, t_3, ...t_{n-1}, and (2) the location of the element $B(t_1)$, $B(t_2)$, $B(t_3)$, ..., $B(t_{n-1})$ at each of these time points, then, of all of the information that we have, the only information that is of use to us is the most recent, that is, t_{n-1} and $B(t_{n-1})$. In predicting the future location of a Brownian element it is not past trajectories that are predictive. Only the most recent past time and position at that time are helpful.*

4.5.3 The Reflection Principle

If we are to apply the tool of Brownian motion to the statistical monitoring of clinical trials, we will need to implement this process' principles in order to compute paths of interest. If, for example, an investigator has a function of a treatment effect that follows Brownian motion, she can compute the probability that this element will exceed the value one at time $t = 2$, as

$$\boldsymbol{P}\left[B(2) \geq 1\right] = \boldsymbol{P}\left[N(0,2) \geq 1\right] = \boldsymbol{P}\left[N(0,1) \geq \frac{1}{\sqrt{2}}\right] = 0.240.$$

This computes the relative frequency of all paths that start at location zero at time $t = 0$ and, two time units later, have a location that is at least one unit away from the origin. However, upon reflection, she may decide that this is not the primary event of interest to her. This computation considers only those paths, that, regardless of the location of the particle at any time point less than two, had a value of at least one at time point 2. It does not consider, for example, those paths that produced locations exceeding the value one at any time between zero and two but were not greater than one at time 2. On consideration, she may be interested in computing the probability of these paths as well.

This second consideration requires us to develop the idea of an element's maximum location. We wish to consider the frequency of paths that attain a maximum location that is greater than one at any time point t such that $0 < t \leq 2$. This problem, admittedly appearing complicated at first glance, has a simple solution, although an exact proof of the solution is beyond the scope of this discussion. Consider the time t when $B(t)$ first reaches the value of one. Then, after that time, there are two possible paths, $B(t)$ and $1-\left[B(t)-1\right]$, which are the mirror image of each other (Figure 4.15).

Each of these paths has the same probability of occurrence. This is the *reflection principle.* Consideration of this observation leads to the conclusion that

$$\boldsymbol{P}\left[\max_{0<t\leq 2} B(t) \geq 1\right] = 2\boldsymbol{P}\left[B(2) \geq 1\right] = 2(0.240) = 0.480.$$

The probability that the maximum excursion of $B(t)$ exceeds a value a during a time period $0 < t \leq T$ is simply twice the probability that the location at the end time T is greater than the value of interest. This is a result that will be of use to us as we develop monitoring rules for clinical research.

---

*A less confusing way to discuss the "memoryless" property of Brownian motion is to say that it is "path-memoryless". However, this is not the tradition, so we will not use this alternate terminology.

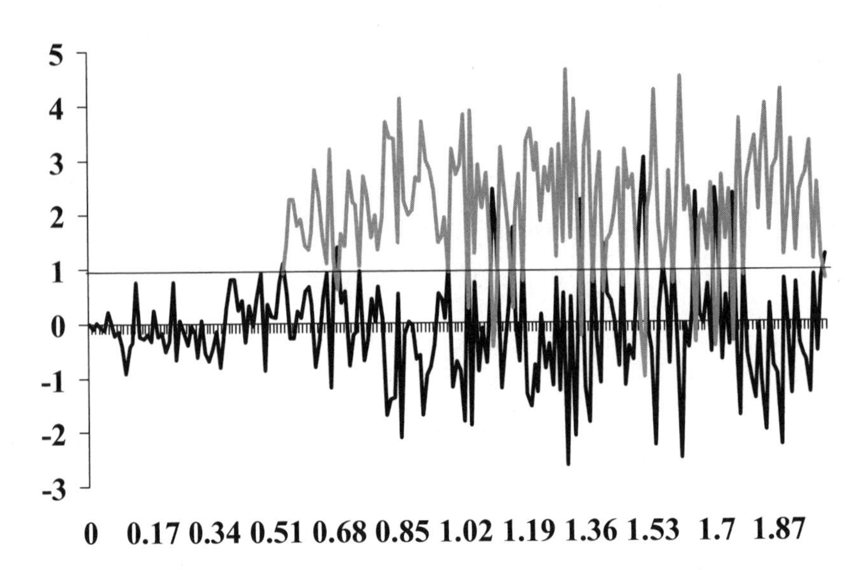

Figure 4.15. Reflection around the line $B(t) = 1$ beginning at the first point where $B(t) >= 1$.

4.5.4 Brownian Motion with Drift

A review of our discussion up to this point (it is hoped, not a memoryless process), reveals that we are working to develop the concept of Brownian motion into a useful implement for monitoring ongoing clinical research. In fact, in our mind's eye, it is becoming increasingly easy to see the similarities between a clinical effect of interest in research and a pollen grain. Both are moving first in one direction, then in another over time.

This analogy is useful when there is no overall treatment effect. However, because the purpose in almost all of clinical research is to identify an effect of interest, all useful monitoring tools must be able to effectively identify this type of result as rapidly as prudent. We therefore, have to consider how our view of the Brownian element's movement changes when there is a systematic influence on its location.

Our understanding of Brownian motion discussed thus far has assumed that the tendency for the element to move in a positive direction is counterbalanced by an equal force that pushes it to the negative. However, what would happen if the element were caught up in a strong steady current upward? In this circumstance, the particle would be swept along by the current, moving in an upward direction. Random forces would still be in operation. However, whereas before these random forces were the only tendencies governing the particle's position, now there is an additional systematic pressure operating to produce movement in one direction.

This systematic component is called *drift*. Brownian motion with drift has the same hallmarks of standard Brownian motion. The probability distribution of the element's location remains normal with increasing variance. Furthermore, the memoryless property of the process remains intact. However, the mean location of the element is different under a drift assumption from that of standard Brownian motion. Under standard Brownian motion, the element's mean location is zero. If the element is drifting, its mean location is μt where μ is the *drift parameter*. This is the only mathematical difference between standard Brownian motion and Brownian motion with drift (Figure 4.16).

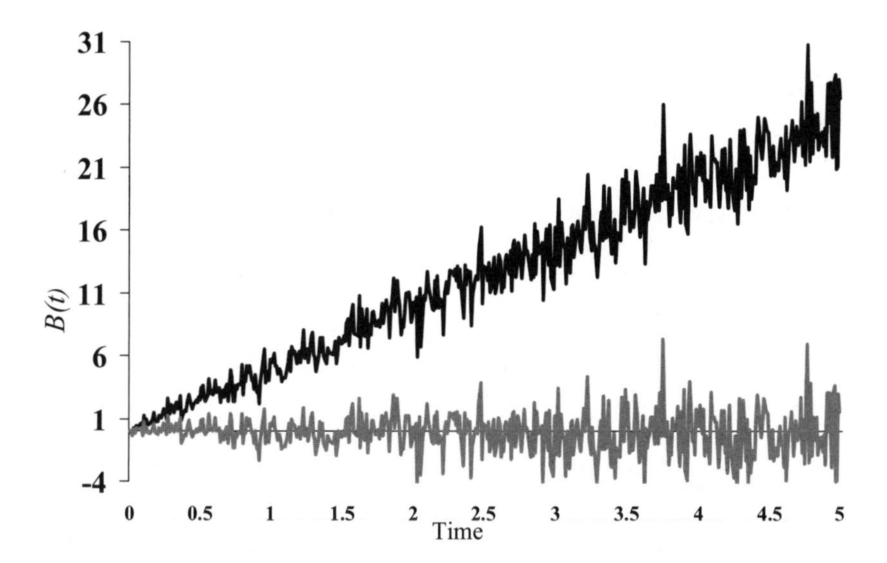

Figure 4.16. Brownian motion with positive drift (black) and with drift removed (gray).

Examination of Figure 4.17 reveals that probabilities for path locations will be different for an element undergoing Brownian motion with drift when compared to standard Brownian motion. As we have seen earlier, if $B(t)$ is following standard Brownian motion, then

$$\boldsymbol{P}\left[B(t) > a\right] = \boldsymbol{P}\left[N(0,t) > a\right] = \boldsymbol{P}\left[N(0,1) > \frac{a}{\sqrt{t}}\right] = 1 - F_Z\left(\frac{a}{\sqrt{t}}\right).$$

However, if the element is undergoing drift with drift parameter μ, then the computation is

$$\boldsymbol{P}\left[B(t) > a\right] = \boldsymbol{P}\left[N(\mu t,t) > a\right] = \boldsymbol{P}\left[N(0,1) > \frac{a-\mu t}{\sqrt{t}}\right] = 1 - F_Z\left(\frac{a-ut}{\sqrt{t}}\right).$$

For example, the probability that an element undergoing standard Brownian motion is greater than the value 1 at time $t = 1$ is $1 - F_Z\left(\frac{1}{\sqrt{1}}\right) = 0.159$. However, if the drift parameter is $\mu = 1$, then this probability becomes $1 - F_Z\left(\frac{1-(1)(1)}{\sqrt{1}}\right) = 0.50$. The probability that an element "drifting" upward will exceed a value is greater than the probability that standard Brownian motion that is experiencing no drift will exceed that value. The notion of Brownian motion will be useful as we consider monitoring clinical research under the assumption that there is an underlying treatment effect.

4.6 Monitoring Research and Brownian Motion

Monitoring clinical research over time has special quantitative needs, and it is only natural that we reach out to the standard statistical tools that have been so reliable in the analysis of clinical research. However, the demands of clinical research monitoring reveal that we will need more than the usual support that comes from statistical measures. Concerns about sampling error inflation demonstrated that repeated testing is not going to be a sufficient protection against the occurrence of large sampling errors. In addition, the fact that our repeated evaluations are dependent upon each other begs the question of how to reflect these interrelationships into any analysis that we execute. We will need special tools to effectively deal with the demands of a rigorous clinical research monitoring program.

A review of Brownian motion reveals that the properties of this type of movement closely align with the monitoring procedure. The fact that Brownian motion follows a normal distribution matches nicely with the normality assumption that accompanies many of the analyzes that we carry out in clinical research. The dependence of the location of a Brownian element on its previous position matches with our assessment of the behavior of a treatment effect that is built on accumulating data over time. In addition, there is a one-to-one correspondence between the notion of drift and the presence of a nonnull effect in a research effort.

The link between the properties of Brownian motion and the challenges that are raised by treatment effects over time would seem to be a direct and useful one. Therefore the focus of the next chapters is on the use of clinical trial monitoring procedures that have their basis in Brownian motion.

Problems

1. What is the new generalization issue of sample-based research that is introduced by the use of interim monitoring procedures?

2. What are the three statistical issues that are raised by the development of statistical monitoring tools?

3. An investigator attempts to monitor her ongoing research effort using the following tool. She generates a test statistic at 25%, 50%, 75%, and 100% of the

time in the study. She will reject the null hypothesis when the p-value of any of these tests is less than 0.001. Assuming that the evaluations are independent, what is the familywise type I error rate? What do you think of the utility of this approach?

4. Consider an investigator who is interested in monitoring his research during its conduct. He plans to assess the value of the test statistic using the following procedure

Percent Time Elapsed in the Trial	Two-Sided P-value
0.20	0.001
0.40	0.005
0.60	0.008
0.80	0.010

Assuming each of these evaluations is independent of the other, show that the type I error level remaining for the final hypothesis test at the end of the study is 0.027, if the overall alpha error rate is to be 0.05.

5. Consider an investigator who is interested in monitoring his research during its conduct. He plans to assess the value of the test statistic using the following procedure

Percent Time Elapsed in the Trial	Two-Sided P-value
0.30	
0.50	0.004
0.75	0.005
0.90	0.006
1.00	0.035

Show that the maximum type I error rate for the first evaluation point of the trial is 0.001 assuming that the tests are to be computed independently and the total alpha to be expended is 0.05.

6. An investigator is monitoring a clinical trial using a function of the effect size that follows Brownian motion with no drift. In this circumstance, the time in the study is measured as percent of time that has elapsed. Compute the probability that this Brownian element exceeds 1.96 when
   a. 5% of the follow-up time of the study has elapsed.
   b. 15% of the follow-up time of the study has elapsed.
   c. 40% of the follow-up time of the study has elapsed.
   d. 75% of the follow-up time of the study has elapsed.
   e. 100% of the follow-up time of the study has elapsed.

7. An investigator is monitoring a clinical trial using a function of the effect size that follows Brownian motion with a drift parameter of $\mu = 1$. In this circum-

stance, the time in the study is measured as percent of time that has elapsed. Compute the probability that this Brownian element exceeds 1.96 when

   a. 5% of the follow-up time of the study has elapsed.
   b. 15% of the follow-up time of the study has elapsed.
   c. 40% of the follow-up time of the study has elapsed.
   d. 75% of the follow-up time of the study has elapsed.
   e. 100% of the follow-up time of the study has elapsed.

8. Investigators are working to identify the circumstances under which a clinical study may be stopped early. They will be monitoring a function of the effect size that follows Brownian motion, and will monitor the clinical study when 40%, 80%, and 90% of the duration of the trial have elapsed. Assuming the drift parameter is $\mu = 0.75$, the investigators are interested in stopping the study when the value of Brownian motion exceeds 3.0 at 40%, 2.5 at 80%, and 2.2 at 90%. Compute
   a. The probability that each of these boundaries is exceeded given that the Brownian element has no drift.
   b. The probability that each of these boundaries is exceeded when the drift parameter is $\mu = 0.75$.
   c. The probability that the Brownian element will exceed 1.96 at the end of the study (i.e., when the follow-up time is 100%) when each of the boundaries is exceeded taken one at a time under the assumption of no drift.
   d. The probability that the Brownian element will exceed 1.96 at the end of the study (i.e., when the follow-up time is 100%) when each of the boundaries is exceeded taken one at a time under the assumption of a drift parameter of 1.96.

References

1. Feller W. (1966) *An Introduction to Probability Theory and its Applications. Volume II.* New York, Wiley. Pages 16–17.
2. Friedman L, Furberg C, Demets D. (1996) *Fundamentals of Clinical Trials.* 3rd Edition. New York, Springer.
3. Piantadosi S. (1997). *Clinical Trials: A Methodologic Perspective.* New York, John Wiley.
4. Moyé L. (2003) *Multiple Analyzes in Clinical Trials: Fundamentals for Investigators.* New York. Springer.
5. Fleming TR, Green SJ, Harrington DP. (1994) Considerations for monitoring and evaluating treatment effects in clinical trials. *Controlled Clinical Trials* **5**:55–66.
6. Hochberg Y, Tamhane AC. (1987). *Multiple Comparison Procedures*, New York, Wiley.
7. Westfall PH, Young SS. (1993) *Resampling Based Multiple Testing: Examples and Methods for P-Value Adjustment.* New York. Wiley.

8. Armitage P, McPherson CK, Rowe BC. (1969) Repeated significance tests on accumulating data. *Journal of the Royal Statistical Society A* **132**:235–244.
9. Mills JL. (1993) Data torturing. *New England Journal of Medicine* **329**:1196–1199.
10. Karlin S, Taylor HM. (1975). *A First Course in Stochastic Processes. 2nd Edition*. New York, Academic.

5

Group Sequential Analysis Procedures

Give us the tools, and we will finish the job

Winston Churchill

Having now completed a brief review of the motivation for monitoring guidelines, and an examination of the building blocks needed to understand the background mathematics of tools to monitor efficacy, we are now prepared to demonstrate and explain their use in modern clinical trial practice.

5.1 Sounding the Alarm

Ethical imperatives motivate the use of statistical monitoring procedures. The proper use of these devices protects patients whose future recruitment to the study may be obviated by the study's early termination. In addition, monitoring tools help to safeguard patients by minimizing their exposure to unhelpful and sometimes harmful interventions in clinical research. However, although we concentrate on the development and use of these monitoring devices, we must also remember that monitoring is more than simple mathematics.

In clinical monitoring, the subjective can sometimes overshadow the objective. These subjective, commonly nonquantitative, considerations include a qualitative assessment of the uniformity of the findings across all study participants. An examination of the occurrence of expected and unexpected adverse events must also be assessed.* Typically, the findings of other independent, sometimes parallel, research efforts are available; if so, these results must also be folded into the termination deliberations. In addition, the biological plausibility of the efficacy response that motivated the termination discussion must be reviewed and, if possible, understood.

The role of mathematics in the decision process to terminate a study is much like that of a vigilant security guard or soldier. It is his job to "sound the

* This is the topic of Chapter Seven.

alarm." Carefully trained to be on the lookout, he identifies and transmits preliminary, but potentially important, findings. However, the response to these warnings should not be reflexive. Deliberations based on the alarm must be carefully considered, proceeding only when all relevant facts and decision implications are well illuminated and in full view.

In clinical trials, the mathematical monitory announced by a "stopping rule" is simply preliminary. Ultimately, its computations may be hastened, incorporated, or overruled by other, more important, human imperatives. This critical acknowledgment will be the foundation upon which we build our understanding of modern monitoring guidelines in clinical research.

5.2 Group Sequential Defined

In the 1980s a set of procedures was developed that have become widely used and accepted in clinical research. They have come to be known as *group sequential procedures*.

The term "group sequential procedure" entered the clinical research lexicon in the late 1970s [1]. It now appears frequently in research protocols, peer reviewed manuscripts, and at IRB and DMC deliberations. However, as is true with many commonly used phrases that have their genesis in a technical field, the vernacular use of the term has blurred its meaning. Our discussion must therefore begin with a definition of the term "group sequential procedure".

Group sequential procedures are simply processes that analyze groups of patients sequentially. At the conclusion of a study, data on all recruited patients becomes available for analysis. This is, of course, not the case during the interim study period. The first interim examination includes patients who were among the first recruited. Patients who are subsequently recruited have their data incorporated into a subsequent group for examination. Thus, the groups are examined in the order in which their data becomes available, i.e., sequentially.

The genesis of the term "group sequential" itself is historical. Wald [2] and Armitage [3] developed the idea of sequential testing in clinical research.* However, at that time, the term applied solely to the pair-by-pair analysis of patients; two patients were recruited, one per treatment group, and had their data analyzed. If the trial was permitted to continue, this same procedure was executed for the next pair of patients. Thus, pairs of patients were evaluated sequentially.

Work in the 1960s expanded this idea. The notion of sequential testing was retained. However, the concept of evaluating patients in pairs was impractical as clinical research designs evolved. Specifically, in the newer protocols (1) consecutive patients were not always in alternating treatment groups,† and (2) the outcome of interest that would determine the effectiveness of the treatment would not be known for months or even years after they were recruited into the study. Thus, the

---

* Discussed in Chapter One.

† The block size is the collection of patients in which the therapy assignment is balanced. If the block size is two, then a pair of patients is divided into one treatment and one control patient. However, larger block sizes (e.g., a block size of four) might result in treatment assignment allocation such that the assigned therapy actually may not alternate from patient to patient.

analysis plan of sequenced pairs was replaced by an analysis plan that focused on sequenced groups of patients (Figure 5.1).

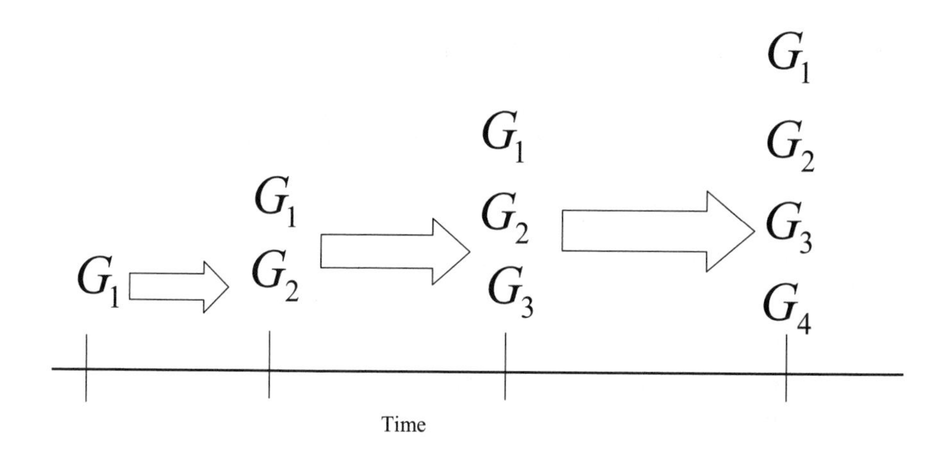

Figure 5.1. The process of group sequential analysis. Groups of patient data arrive chronologically and are combined into an accumulating data set for evaluation.

In the group sequential procedure, each group's data is added to the data that has been collected and is already available from the previous groups. This process generates a growing body of knowledge that sheds light on the emergence of efficacy in the study. The utility of analysis plans based on grouped sequenced data was recognized by Pocock [1].

5.3 Group Sequenced Data and Brownian Motion

The concept of the group sequential analysis makes good sense, but begs the question as to how this incoming grouped data should be analyzed. Fortunately, we are now in a position to address the salient issues raised by the concept of group sequential analysis.

Chapter Four demonstrated how accumulating data that is gathered a group at time, with each group's data added to a growing dataset, represents a body of information that contains built-in dependencies. Thus, the optimal analysis would be one that takes this dependency into account. An advantage of this approach would be to minimize the inflation of the type I error rate as the growing dataset is sequentially analyzed.

However, the nature of the dependence i.e., exactly how a previous examination of early data illuminates our analysis of current data must be elucidated in order for the investigator to make the appropriate adjustments. The important work of unraveling this dependency was carried out by Slud and Wei [4]. Their efforts demonstrated that measures of efficacy in clinical research that were derived from group sequential data could be transformed into an element that followed Brownian motion. This finding permits us to apply much of the understanding that we have about Brownian motion to the analysis of incoming data.

In group sequential procedures, we commonly need to convert the measure of efficacy that the investigator is using (commonly a test statistic) into a statistic that follows Brownian motion. We will call this transformed efficacy measurement a *Brownian element*, or *monitoring device*. This, we will see later, is a very easy step. We will then be in a position to translate a statement in probability about the clinical efficacy measure into an equivalent statement about the Brownian element. Because the events are equivalent, an answer to this latter question will provide the required solution to the first question about the level of therapy effectiveness.

We may think of this entire process as a five step operation to address the question, "Is there sufficient measure of efficacy at this point in the study to consider its early termination?"

Step 1. Have in hand a measure of efficacy at a point in the study
Step 2. Convert the question of early efficacy into a probability question.
Step 3. Convert this measure of efficacy into a Brownian element and construct an equivalent probability question about the Brownian element.
Step 4. Answer the probability question about the Brownian element.
Step 5. Translate this answer back to the answer for the original question about the level of clinical efficacy.

Step 1 merely requires that the investigator have a test statistic at the interim monitoring point. Once we accomplish step 2, then steps 3 through 5 are relatively easy to take.

5.4 Analysis Classes for Group Sequenced Data

The next important issue is the use of the incoming, group sequenced data to address whether the research should be discontinued early because of an early finding of efficacy. There are, in general, two accepted approaches to the examination of this question.

5.4.1 Looking Backward

Assume that an investigator has a measure of efficacy at an interim point in the study. She wishes to determine if its value suggests that the study should be terminated early. One perspective would be to infer something about the pattern or trajectory of the efficacy measure; if the path that it followed from the beginning of the study was an atypical one under the assumption that there was no efficacy, then its unusual track may suggest that the underlying null hypothesis of no efficacy was wrong. Therefore, a natural procedure that she might use would be to look backward to see if the path that the statistic has followed was particularly unanticipated. If its observed trajectory deviates substantially from the expected trajectory under the assumption of no efficacy, then the investigator would be inclined to believe that some consideration should be given to terminating the study (Figure 5.2).

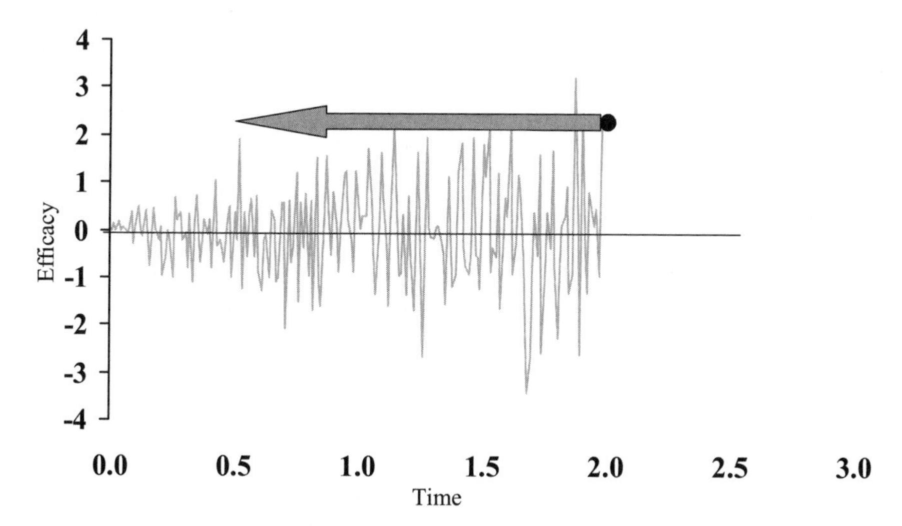

Figure 5.2. First group sequential procedures looked backward. The question asked is, "How likely is the path (under the null hypothesis) of the test statistic that led to its present position?" The path (grey) is typically not known. The only known quantity is the test statistic (in black).

In Figure 5.2, the current value of the monitoring statistic is noted by the black ball. With this quantity in hand, the investigator looks back to the beginning of the study, asking how likely is it that the monitoring statistic would have the value it currently has if there were no substantial early emergence of clinical effectiveness in the research. If the monitoring statistic was very likely to follow the observed trajectory under the null hypothesis of no treatment effect, then the observed value of the monitoring statistic is consistent with the hypothesis of no therapy effect; this suggests that nothing out of the ordinary has occurred, and that the trial should be allowed to continue. This "look backward" approach is the first group sequential procedure that came into general acceptance in the 1970's.

It is important to note that the investigator does not know the details of the daily oscillations of the monitoring device over time. She only knows the value at its current point and the value at the origin. Because many paths could occur that would lead to the current value of the monitoring device, she must compute the relatively frequency of all of these paths under the null hypothesis. She cannot do this directly if she remains focused on the clinical measure of efficacy, but she can accomplish it if she translates the process to one involving a Brownian element.

5.4.2 Looking Forward

A second collection of procedures addresses the question of early efficacy from a different perspective. Rather than take a look backward over the path that has been traversed, the investigator looks forward to the end of the study (Figure 5.3). In this scenario, the investigator with the monitoring device in hand, asks, "How likely is it that the test statistic will fall in the critical region at the end of the study?" If this is a very likely scenario, then the investigator may be tempted to end the study early. This forward-looking approach is one that projects the future path of the monitoring statistic.

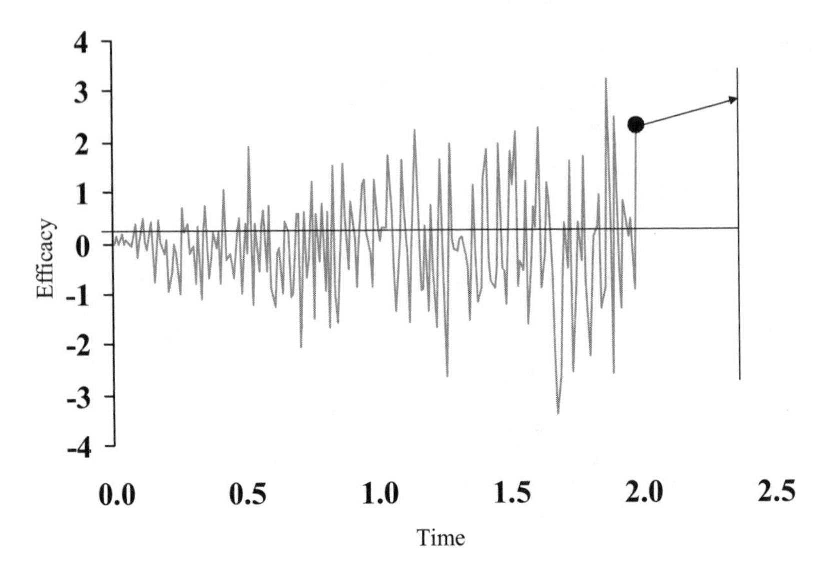

Figure 5.3 A look forward approach. This perspective assesses the likelihood that the test statistic will lead to rejection of the null hypothesis at the conclusion of the study.

Both the backward look and the forward look are group sequential procedures, because they each examine data that are accumulated in "groups" over time. Both sets of procedures use the concept of Brownian motion in order to assess the relevant probabilities for the investigator. In this chapter we will focus on the procedures that look backwards. The forward look procedure, more commonly described as a *conditional power* or *stochastic curtailment* [5], will be the focus of Chapter Six.

5.5 Analysis Complexities

Group sequential procedures are natural and intuitive perspectives to develop. The statistical procedures that have been developed to support these perspectives are, in principle, elementary, and in some circumstances, are straightforward (Figure 5.4).

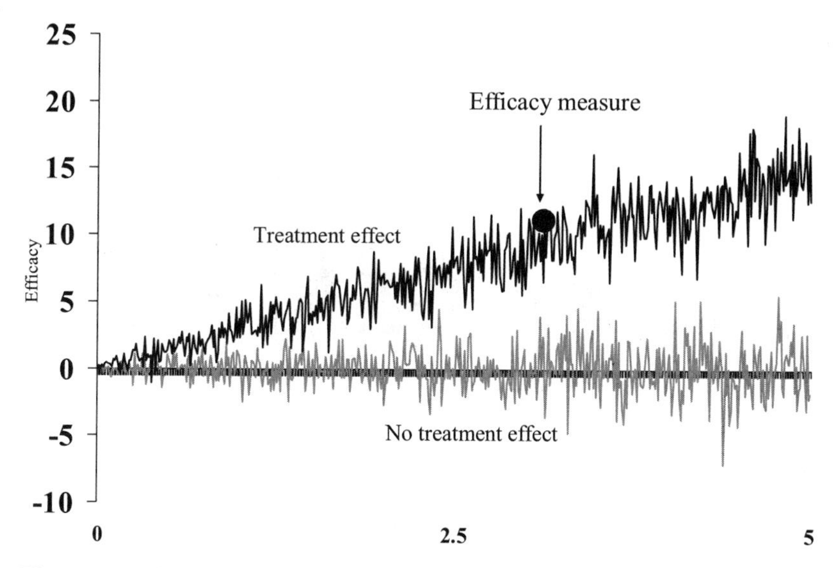

Figure 5.4. In this case, the measure of efficacy is clearly consistent with a treatment effect.

In Figure 5.4 the measure of efficacy from a Brownian element is plotted over time. Two scenarios are presented, each represented by a single path. The lower path reflects the movement of the Brownian element over time in the clinical circumstance when the therapy measure is not effective. In this scenario, the efficacy measure moves both above and below zero. As we have come to expect, the monitoring device measure has excursions of increasing magnitude from zero as time progresses. However, over the course of time, the measure demonstrates no persistent tendency to remain either above or below zero. The average location of the point is close to zero, although the number of large excursions away from zero increases.

The second scenario from Figure 5.4 demonstrates a treatment positive path. In this circumstance, there is important consistent motion of the efficacy measure away from zero. Although there are both positive and negative excursions, the number of positive movements exceeds the negative ones. The Brownian element is consistently lifted above zero. In this case, the efficacy measure is clearly embedded in the efficacy path.

In the scenario presented in Figure 5.4, any statistical procedure that is chosen will have a relatively easy task of differentiating between the two depicted paths, because they clearly separate. Even in this case, however, the earlier the research program is evaluated, the more difficult it becomes to discern on which particular path the efficacy point lies.

However, in other circumstances, the results are not so clear cut. Consider the monitoring scenario that occurs when the treatment effect is not considered to be as dramatic as was observed in the previous example. In this new scenario it can be very difficult to separate the "no-effect" path from that of a treatment effect (Figure 5.5).

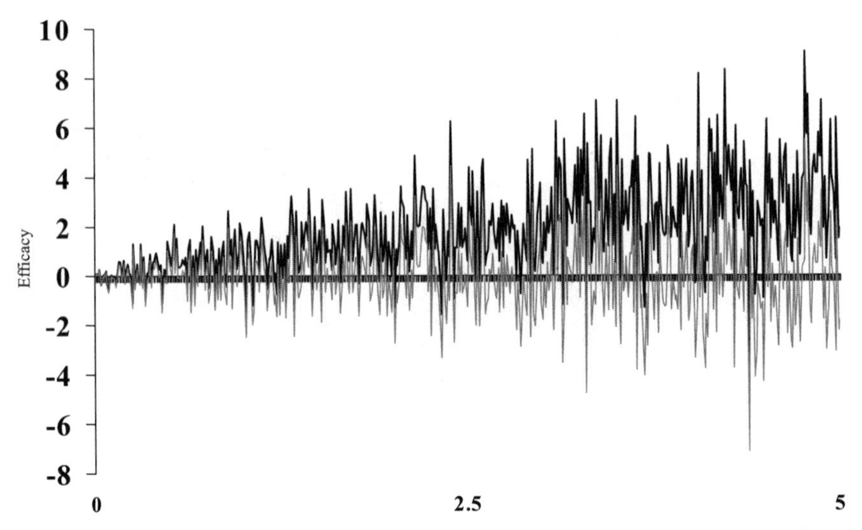

Figure 5.5. It can be difficult to distinguish between a borderline beneficial treatment effect path (dark) from a path of no effect (gray).

In this setting, there is substantial overlap between the efficacy path when the efficacy is small versus when there is no efficacy whatsoever. Any statistical procedure that we can envision will have difficulty in discriminating between these two scenarios.

Examples of this scenario are clinical studies in which the effect of the therapy that is being tested takes years to appear. One such illustration is the CARE study. This clinical trial evaluated the effect of the HMG-CoA reductase inhibitor pravastatin on the five year cumulative incidence rate of fatal and/or nonfatal myocardial infarction [6]. In this circumstance, it took two years of follow-up before the signal of treatment efficacy emerged from the background, sampling error noise.*

5.6 Using Brownian Motion to Monitor Efficacy

We pointed out in the previous section that one intuitive way to assess which path is best represented by the small number of available efficacy measurements is to review the trajectory that has been traversed thus far. The question addressed by this look-back approach is, given the data that have been observed, how likely is it that

* One might argue that the dilemma of attempting to separate a signal that takes at least two years to develop (as was the case in CARE) be obviated by simply not monitoring the primary analysis of the study for two years. However, one of the goals of interim monitoring is to be vigilant for the unexpected, a responsibility that requires the DMC to deliberately inject itself into discussions produced from the conundrum represented in Figure 5.5.

the path traveled by the efficacy measure is consistent with the expected path. Consider the following adaptation of Figure 5.4 (Figure 5.6).

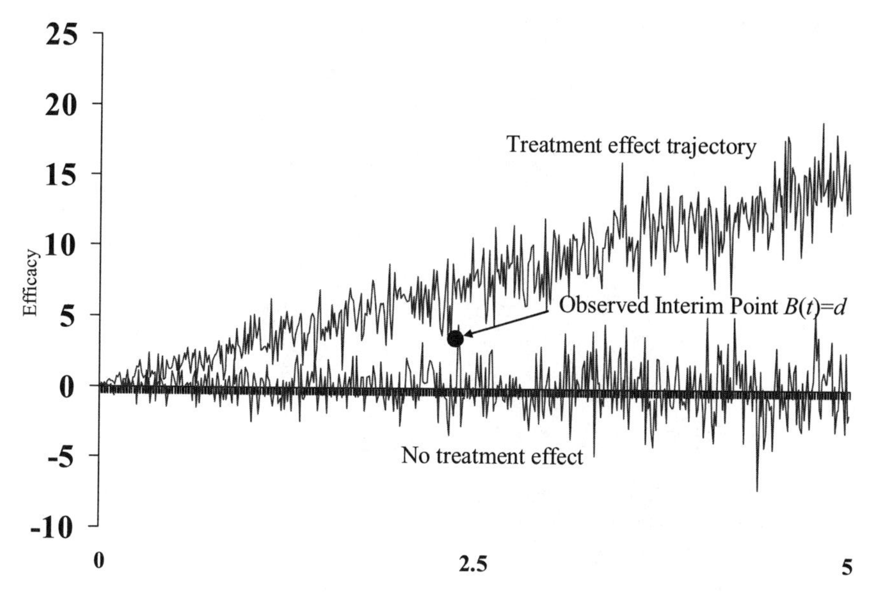

Figure 5.6. Attempting to distinguish a clear treatment effect based on one interim point.

In Figure 5.6, there is one interim measure of efficacy whose value at time t is d. The figure also presents two of the many possible paths for the Brownian element, one indicative of no treatment effect, the second demonstrating a positive treatment effect path. The task of the investigator is to determine which of these paths is most consistent with the interim measure of efficacy seen to have the value d.

5.6.1 Introductory Computations

Recall that the look-back procedure simply asks how likely is the path that the monitoring device has taken up to the current point. However, in order to answer this question we must clarify a condition. Is this question being asked under the null hypothesis of no therapy effect, or under the alternative hypothesis in which an effect is present? This condition will have a critical effect on our assessment of the likelihood of the monitoring statistic's path.

If for the moment we work under the null hypothesis, then the question might be phrased as, "How likely would an interim value of the monitoring statistic d be at time t be under the assumption of no treatment effect," However, recall that there are uncountably many possible values for the Brownian element at time t (Figure 5.7).

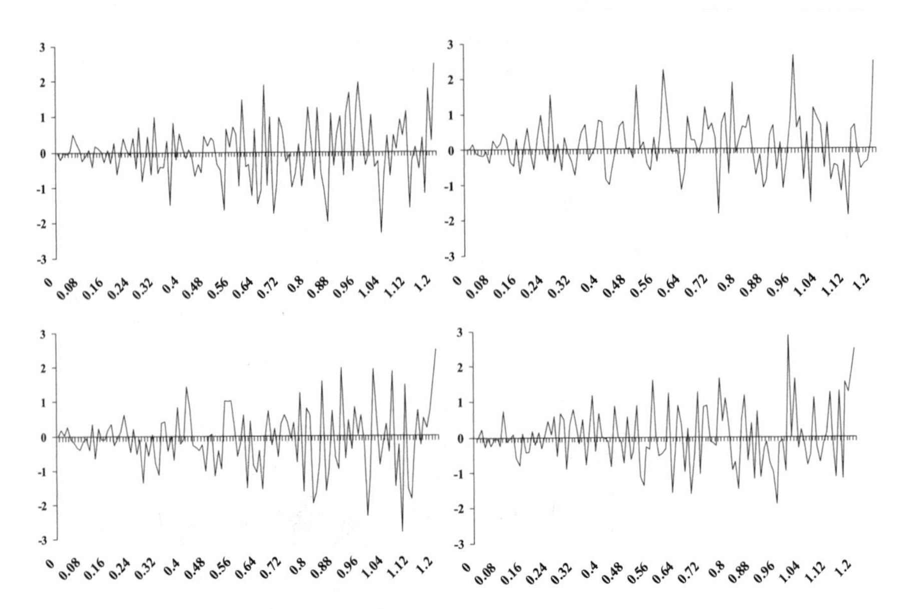

Figure 5.7. Four different paths that have a value of 2.49 at time 1.20. The look backwards procedure computes the probability of these four (and all other) paths that have a value at least as large as 2.49 at time 1.20.

Because the probability of any one of them is zero, we therefore must rephrase the question to "How likely are all of the paths that produce a monitoring statistic with a value at least as large as d at time t under the assumption of no treatment effect?" This permits us to use the probability as area concept from Chapter 3. Because we are following a Brownian element, the solution is easy.

$$\boldsymbol{P}\left[B(t) \geq d\right] = \boldsymbol{P}\left[N(0,t) \geq d\right] = \boldsymbol{P}\left[N(0,1) \geq \frac{d}{\sqrt{t}}\right] = 1 - F_Z\left[\frac{d}{\sqrt{t}}\right], \quad (5.1)$$

where $N(0,1)$ refers to a variable that follows a normal distribution with mean zero and variance one. This is a result that we saw from Chapter Four. The assumption of no treatment effect permits us to work under the assumption of standard Brownian motion. If the value from expression (5.1) is large, then the interim observation of the Brownian element being equal to or exceeding d is quite likely under the null hypothesis of no treatment effect. Because a noneffective therapy would be expected to provide paths containing the point d, we do not have evidence that the therapy is exerting an effect merely because the path arrived at point d.

However, the question of the current path's likelihood could also be addressed under the alternative hypothesis. In this setting, there is a treatment effect. This new assumption is equivalent to assuming that the Brownian element has a drift associated with it. Recall from Chapter Four that this drift parameter was designated as μ. Thus, in order to address the question of how likely is the observed path assuming a treatment effect, we write

$$\boldsymbol{P}\left[B(t) \geq d\right] = \boldsymbol{P}\left[N(\mu t, t) \geq d\right] = \boldsymbol{P}\left[N(0,1) \geq \frac{d - \mu t}{\sqrt{t}}\right] = 1 - F_Z\left[\frac{d - \mu t}{\sqrt{t}}\right]. \quad (5.2)$$

As an example, consider a clinical experiment in which at time $t = 2$, $d = 5$. Then we can compute under the null hypothesis:

$$\boldsymbol{P}\left[B(2) \geq 5\right] = \boldsymbol{P}\left[N(0,2) \geq 5\right] = \boldsymbol{P}\left[N(0,1) \geq \frac{5}{\sqrt{3}}\right]$$
$$= 1 - F_Z\left[\frac{5}{\sqrt{3}}\right] = 1 - F_Z\left[2.89\right] = 0.002.$$

This suggests that a path that is at least as extreme as having a value of 5 at time $t =$ 2 is very unlikely. However, that same path, when considered from the alternative where the drift parameter $\mu = 3$ is much more likely. We compute using equation (5.2) that

$$\boldsymbol{P}\left[B(3) \geq 5\right] = \boldsymbol{P}\left[N((3)(3),\, 3) \geq 5\right] = \boldsymbol{P}\left[N(0,1) \geq \frac{5-9}{\sqrt{3}}\right] = 1 - F_Z\left[\frac{5-9}{\sqrt{3}}\right]$$
$$= 1 - F_Z\left[-2.31\right] = 0.990.$$

Thus, this same path for the test statistic is very much more likely under the alternative hypothesis than under the null hypothesis, and we would be very tempted to reject the null hypothesis in favor of the alternative assumption of a treatment effect.

Of course the complexity of the monitoring problem is compounded when one has only a small number of points (i.e., three or four) on which to decide whether it is likely that an early treatment effect is emerging. The difficulty that is raised with only having a relatively small number of locations on the efficacy path is that these points can be consistent with many (in fact, uncountably many) paths. The analysis task before us is to identify the class or category of points most consistent with the data that has been collected thus far.

5.6.2 Information Time

It is perhaps a truism that, if all of the information were available immediately and at once in clinical research, then there would be no need for monitoring. Therefore, monitoring clinical research for efficacy relies on a fundamental operational principle of clinical studies: information on efficacy becomes available over the course of time.

However, the mere passage of time is not always synonymous with accrual of information, i.e., clinical events. For example, consider the monitoring design for a clinical study. The randomized, controlled clinical trial is designed to assess the effect of therapy on the total mortality rate at one year. The sample size for the study is computed to be 1200 patients. Based on the anticipated mortality rate that these patients will experience, the investigators anticipate that at the one-year conclusion of the study, there will be 100 total deaths in the trial.

If the investigators wish to monitor the study at the 50% time point, then what proportion does 50% reflect? A natural but wrong answer might be that 50% simply identifies that point when 50% of the follow-up time has elapsed, that is, 50% measures the passage of time. However, commonly deaths do not occur uniformly over time. Sometimes they occur earlier than anticipated, whereas in other circumstances the deaths are late events. Occasionally, the investigators actually overestimate the death rate so that it takes longer than one year to accrue the 100 deaths that the investigators would need to conclude the study.

Thus, the pertinent contribution to the monitoring device is not the passage of time, but the accumulation of deaths. This realization is the motivation for the term *information time.*[*] The information time is simply the fraction of the total number of endpoint events being monitored that have been collected at a given point in the follow-up period of the clinical trial. In this example, what provides ballast to the interim efficacy analysis is not the passage of time, but the number of deaths.

Therefore, in group sequential procedures, information time replaces actual time in monitoring clinical events. The use of information time severs the link between the passage of real time, which is coincidental to the monitoring process, and the accrual of clinical events which is central to the monitoring procedure.

5.6.3 Monitoring Device and Brownian Motion

Implementing this concept of information time permits us to link a device or statistic that follows Brownian motion to the original test statistic. Our goal now is to identify a function of the monitored data that follows Brownian motion, and then show how that function is related to the actual test statistic. This will permit us to convert complicated events involving the test statistic into simpler events that follow Brownian motion.

As a first example, consider a very basic research design. An investigator is interested in examining the 24-hour post-surgical mortality rate p for patients who undergo limb amputation. At the conclusion of the study, the investigator will have examined N patients. Under the null hypothesis, this rate p is believed to equal some value p_0 that is available from the literature. The investigator is interested in conducting an interim review of his data when the results are available for only n of the patients, where $0 \leq n < N$. The question to be addressed is what device (i.e., what function of the data) should be used to monitor this process.

Let x_i be the mortality status of the ith patient in the trial. Let $x_i = 1$ if the ith patient dies within 24 hours of surgery, and $x_i = 0$ if this patient survives at least as long as 24 hours. By the end of the study, results will be available for $i = 1, 2, 3, \ldots, N$. In this case the best estimator of the 24-hour mortality rate based on N patients is

---

[*] This is an approximation of Fisher's information, which is related to the inverse of the variance.

$\overline{X} = \sum_{i=1}^{N} x_i \Big/ N$. However, at the interim point of examination of the data, the result is available for only n patients. We turn to what we know about the probability distribution of this clinical event of interest to aid in the construction of the monitoring device.

We know that the mean of x_i is p_0 under the null hypothesis, and its variance is $p_0(1-p_0)$. Invoking the central limit theorem discussed in Chapter Three, we may write that

$$\sum_{i=1}^{n} x_i \sim N\left(np_0, np_0(1-p_0)\right). \tag{5.3}$$

We have seen that if the monitoring statistic that we are working to find is to have a distribution that follows Brownian motion, then its mean and variance must be a function of "information time". Begin by dividing the sum of the x_i's by $\sqrt{N}$ to reveal

$$\frac{\sum_{i=1}^{n} x_i}{\sqrt{N}} \sim N\left(\frac{n}{\sqrt{N}} p_0, \frac{n}{N} p_0(1-p_0)\right). \tag{5.4}$$

Note that the left hand side of expression (5.4) is not the sum of observations divided by the square root of the number of those observations. The number of observations reflected in the numerator is n, the total number available at the interim monitoring time point. However, the number of observations reflected in the denominator is N, the total number that is available in the population.

The changes in the mean and variance of the normal distribution on the right hand side of expression (5.4) are based on two facts: (1) a variable that follows a probability distribution when divided by a constant has its mean divided by that same constant, and (2) its variance is the original variance divided by the square of that constant. We proceed by writing

$$\frac{\sum_{i=1}^{n} x_i}{\sqrt{N}\sqrt{p_0(1-p_0)}} \sim N\left(\frac{n}{\sqrt{N}\sqrt{p_0(1-p_0)}} p_0, \frac{n}{N}\right). \tag{5.5}$$

An examination of the variance of this new variable expressed in (5.5) reveals that it is exactly the proportion of total patients expected by the study's conclusion. This proportion is the information time, and we write $I = n \big/ N$. Note that, at the beginning of the study, there are no patients and $I = 0$. At the conclusion of the study, $n = N$ and $I = 1$. This gradual accrual of information over time is exactly what information time is designed to measure.

If the null hypothesis is true, and the post-surgical mortality rate is equal to p_0, then it is easy to see that the monitoring device

$$\frac{\sum_{i=1}^{n} x_i}{\sqrt{N}\sqrt{p_0(1-p_0)}} - \frac{np_0}{\sqrt{N}\sqrt{p_0(1-p_0)}} = \frac{\sum_{i=1}^{n} x_i - np_0}{\sqrt{N}\sqrt{p_0(1-p_0)}}$$

is normally distributed with mean 0 and variance n/N. Thus, if the null hypothesis H_0: $p = p_0$ then the statistic

$$\frac{\sum_{i=1}^{n} x_i - np_0}{\sqrt{Np_0(1-p_0)}}$$

follows standard Brownian motion. It is this statistic that will be our monitoring device (Figure 5.8).

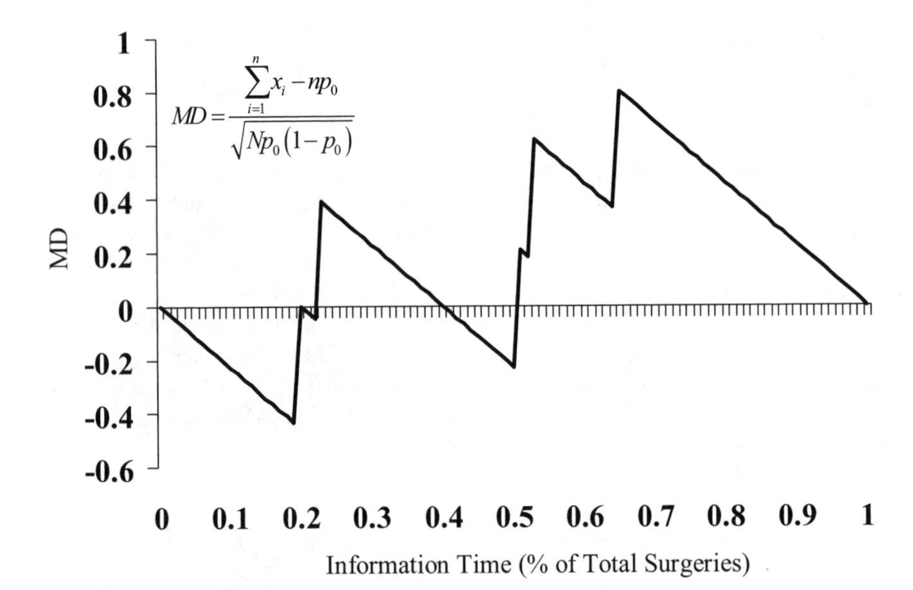

Figure 5.8. Monitoring device (MD) for estimating 24-hour post-surgery mortality rate over information time.

The difficulty with this device is that it is unfamiliar. Being such, the investigators will have some difficulty couching clinical events of interest in terms of this unrecognized statistic. If this device is to be useful, we must be able to convert statements about the measure of effect in the study (in this case, the disparity between the observed and predicted post-surgical mortality rate) into equivalent events that involve the monitoring device. Focusing on the monitoring device *B(I)*, we may write:

$$B(I) = B\left(\frac{n}{N}\right) = \frac{\sum_{i=1}^{n} x_i - np_0}{\sqrt{N}\sqrt{p_0(1-p_0)}} = \frac{n(\bar{x}_n - p_0)}{\sqrt{N}\sqrt{p_0(1-p_0)}}. \qquad (5.6)$$

Multiplying by $\sqrt{N}/\sqrt{n}$ we have

$$\frac{\sqrt{N}}{\sqrt{n}} B\left(\frac{n}{N}\right) = \frac{\sqrt{N}}{\sqrt{n}} \frac{n(\bar{x}_n - p_0)}{\sqrt{N}\sqrt{p_0(1-p_0)}} = \frac{\sqrt{n}(\bar{x}_n - p_0)}{\sqrt{p_0(1-p_0)}} = \frac{(\bar{x}_n - p_0)}{\sqrt{\frac{p_0(1-p_0)}{n}}}. \qquad (5.7)$$

But the expression on the right hand side of equation (5.7) is simply the test statistic for a test of the hypothesis H_0: $p = p_0$ versus H_a: $p \neq p_0$ using a one sample binomial test of proportions. In this circumstance, the test statistic is computed when there are n observations available, or $I = n/N$. We will call this test statistic *TS(I)*. Therefore, we have demonstrated that

$$TS(I) = \sqrt{\frac{N}{n}} B(I) = \frac{B(I)}{\sqrt{I}} \qquad (5.8)$$

or

$$B(I) = \sqrt{I}\, TS(I). \qquad (5.9)$$

This is the statement of equivalence that we need. We are now in position to convert events that involve test statistics into an event that involves Brownian motion.

To show that both the monitoring device (which under the null hypothesis, follows standard Brownian motion) and the test statistic can measure equivalent events, consider the following example. Assume that with 40 subjects in the study, the investigators have observed 5 post-surgical deaths. They would like to compute how likely this event is under the null hypothesis using a look backward approach. With this data in hand, they compute $\bar{x}_n = 5/40 = 0.125$. They calculate that, at this point, $I = 40/100 = 0.40$, and the investigators can compute $TS(0.40)$ as

$$TS(0.40) = \frac{(\bar{x}_n - p_0)}{\sqrt{\frac{p_0(1-p_0)}{n}}} = \frac{(0.125 - 0.05)}{\sqrt{\frac{0.05(1-0.05)}{40}}} = 2.18.$$

If the investigators were then to directly compute a p-value for this test statistic, they would observe that $p = 0.029$, which might suggest to them that the study could be discontinued. The computation considered from the Brownian motion perspective follows. The investigators compute

$$\begin{aligned} P\left[\left|TS(0.40)\right| \geq 2.18\right] &= P\left[\left|\sqrt{0.40}\,TS(0.40)\right| \geq \sqrt{0.40}\,(2.18)\right] \\ &= P\left[\left|B(0.40)\right| \geq 1.379\right]. \end{aligned}$$

Here, we have converted an event involving a test statistic into an event involving Brownian motion. The absolute value simply addresses the concern about both large positive and large negative deviations of the observed from the expected post-surgical mortality rate; this is the equivalent of carrying out a two-tailed test. We can now apply what we know about computing probabilities involving Brownian motion to this circumstance, writing

$$\begin{aligned} P\left[\left| B(0.40) \right| \geq 1.379\right] = P\left[\left| N(0, 0.40) \right| \geq 1.379\right] &= 2P\left[N(0,1) \geq \frac{1.379}{\sqrt{0.40}}\right] \\ &= 2\left(1 - F_Z(2.18)\right) = 0.029. \end{aligned}$$

Thus, in this scenario, the computation of the relevant probability leads to the same result.

However, we have not come all this way to merely compute an easily obtained probability by using a more complex Brownian motion argument, but instead to expand the repertoire of events whose relative frequencies are available. An event of interest to the investigators is, given that the test statistic is equal to 2.18 when 40% of the events are available, what is the probability that it will be greater than 2.18 when 60% of the information is available under the null hypothesis of no treatment effect? This is a computation that will have important implications for us in the next chapter. Recalling the discussion of conditional probability of Chapter Three, we can describe the probability of interest as

$$P\left[TS(0.60) > 1.96 \mid TS(0.40) = 2.18\right]. \tag{5.10}$$

This is a computation that would be difficult to approach using only what we know about test statistics and hypothesis testing. However, its solution is straightforward if we convert this event to one involving Brownian motion. Remembering that $B(I) = \sqrt{I}TS(I)$, we first convert the probability in (5.10) to

$$P\left[\sqrt{0.60}\,TS(0.60) > \sqrt{0.60}\,(1.96) \mid TS(0.40) = 2.18\right].$$

Multiplying each side by $\sqrt{I}$ does not change the event, but simply produces an equivalent event that is now couched in terms of Brownian motion. Applying the same transformation to the condition $TS(0.40) = 2.18$ reveals

$$\boldsymbol{P}\left[\sqrt{0.60}\,TS(0.60) > \sqrt{0.60}\,1.96 \mid \sqrt{0.40}\,TS(0.40) = \sqrt{0.40}\,(\,2.18)\right].$$

We may now write this probability in terms of Brownian motion

$$\boldsymbol{P}\left[B(0.60) > 1.52 \mid B(0.40) = 1.38\right].$$

The conditional probability in terms of the original test statistic has now been converted to an equivalent test statistic for Brownian motion. However, recognizing that this is the probability of an event that is conditioned on the past, we can write

$$\begin{aligned}\boldsymbol{P}\left[B(0.60) > 1.52 \mid B(0.40) = 1.38\right] &= \boldsymbol{P}\left[B(0.60-0.40) > 1.52-1.38\right]\\ &= \boldsymbol{P}\left[B(0.20) > 0.14\right].\end{aligned}$$

The real utility of Brownian motion appears in this setting. The conditional probability that was an obstacle when the event appeared in terms of a test statistic is easily handled when it is rewritten in terms of Brownian motion. We can now quickly finish the computation.

$$\begin{aligned}\boldsymbol{P}\left[B(0.20) > 0.14\right] &= \boldsymbol{P}\left[N(0,0.20) > 0.14\right]\\ &= \boldsymbol{P}\left[N(0,1) > \frac{0.14}{\sqrt{0.20}}\right] = \boldsymbol{P}\left[N(0,1) > 0.313\right] = 0.337.\end{aligned} \tag{5.11}$$

We see that given the test statistic has value of 2.18 when 40% of the information time is available, the probability that the test statistic at 60% of the information time is at least that value is 0.337. We will return to this type of computation when we discuss conditional power in Chapter Six.

5.7 Boundaries and Alpha Spending Functions

The style of computation that was provided in the previous section has been both revealing and interesting. Specifically, the introduction of Brownian motion provides a tool for us to compute the probability of complicated events involving interim measures of efficacy in clinical research. Useful as these computations may be, though, there are two concerns about them that we must confront. The first is that the calculations are not prospective, and the events whose probability that they compute have thus far been data-based. For example, the previous illustration examined the possible location of future values of the test statistic at 60% information time, based on the value of the test statistic when 40% of the information was available. This was a question that was not raised until *TS*(0.40) was recognized, that is, it was a question based on the contents of the incoming data stream. Although the computation was illuminating, interest was not generated in it prospectively. As we saw in Chapter Three, the answers to data-driven questions have important limitations.

Secondly, we have not yet dealt with the accumulating type I error issue that accompanies multiple hypothesis tests on the same dataset. This is relevant here because the investigator can make a decision at $I = 0.40$ to discontinue the trial. Because that decision will be driven in part by the value of the monitoring device, a device which we know to be influenced by sampling error, the influence of this error must be tracked and reported.

The O'Brien–Fleming and Lan–DeMets procedures are based on an adaptation of this approach and allow us to deal with these two important issues. The ubiquity of interim monitoring devices is due in large part to solutions to each of these questions. Specifically, mathematical advances now allow the use of a prospectively declared computation to govern when actions should be considered to terminate the study for efficacy. In addition, calculations can now appropriately accumulate accurate type I error rates over the course of the repeated monitoring assessments.

5.7.1 Boundary Values

The need for a prospectively declared decision rule for considering termination of a study is addressed by the concept of a boundary. This boundary divides the location of the monitoring device into regions. One region suggests that the study can be terminated because sufficient information has been gathered concerning the presence of efficacy. If the monitoring device falls into the other region, then sufficient efficacy has not yet been produced and the research should be permitted to continue. Although the computation of these boundary values can be easily accomplished on almost any computer, the actual calculations are nevertheless complicated. This is especially true of the look-back procedures, as well as some Bayesian paradigms. However, an understanding of how the boundary values are computed, as well as the motivation for the shape of the regions is well within our reach.

Before we begin these discussions, however, we must recall that crossing any such boundary is only one piece of information to consider in terminating a study.

5.7.1.1 Motivating the Shape of the Boundary

As we saw earlier, the boundary of a monitoring statistic serves as a guide to aid us in our reaction to the value of that statistic at an interim point during the execution of the trial. However, the shape of these regions has undergone continued evolution. Initially these regions were simply rectangles, demarcated by straight lines. However, the increased sophistication of mathematical procedures now permits the investigator to consider boundaries that have complex shapes. Each of these shapes can be linked to the underlying assumptions of the monitoring device, and to the particular needs of the trial that is being monitored. The important message for the investigators is that they have a role to play in choosing the shape of the monitoring boundary. The creation of a boundary instantly solves the prospective decision rule (Figure 5.9). This prospectively drawn boundary allows interim monitoring decisions to be in place based only on a priori considerations. This permits us to avoid relying on the appearance of trends in the data that may be due to sampling error. The flat boundary approach was used by Pocock [1].

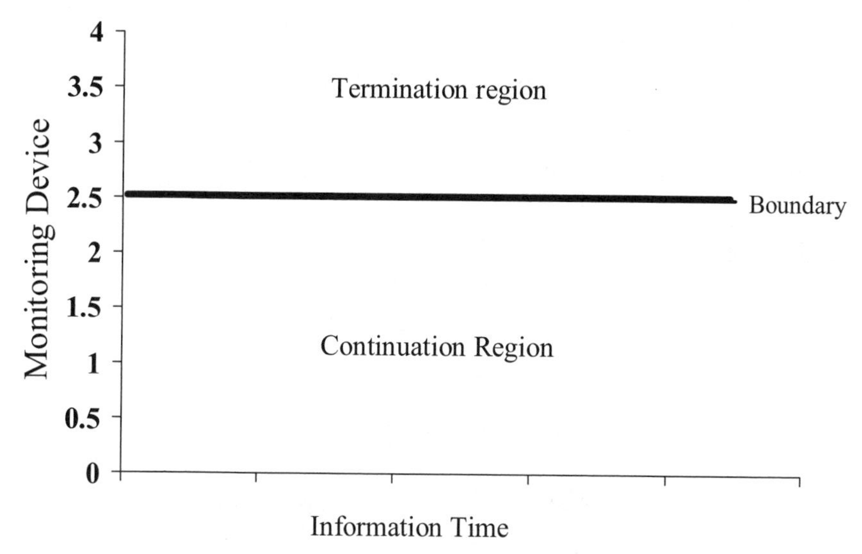

Figure 5.9. A flat line boundary that separates the possible locations of the monitoring device into two regions; (1) a continuation region and (2) a termination region.

However, this flat line of demarcation of Figure 5.9 suggests that the strength of evidence for a research effect is independent of the information time at which the monitoring device is interrogated, an assumption that we now recognize must be carefully considered in clinical circumstances.

An alternative is a boundary that has a positive slope. The class of tests developed to address this circumstance falls under the rubric of the Whitehead Triangular Test [7,8]. An advantage of this procedure is that the computations needed to compute its boundary values are not especially complicated, and its boundaries align closely with those of Wald's sequential probability ratio test [9].

In order to address the concern that early values of the monitoring device provide less stable information because they are based on relatively small datasets, the flat boundary is commonly replaced by the use of the sloping boundary (Figure 5.10).

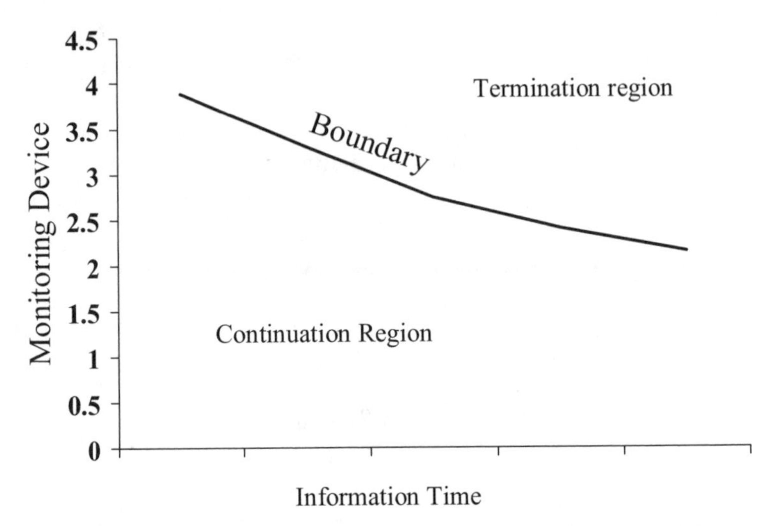

Figure 5.10. The sloping boundary approach. Higher values of the monitoring device are required early in the study to trigger the consideration of terminating the study.

The sloping boundary approach institutes the requirement that the criteria for discontinuing a clinical trial actually change over the course of time. Early in the study the boundary is higher, requiring that the monitoring device itself be larger.

This approach aligns nicely with both the mathematics and our underlying intuition based on the reliability of the monitoring device. Recall that, early in the study, the clinical trial has information on only a small number of subjects, and the number of clinical events on which the value of the monitoring device rests is small. Therefore, the measure of efficacy, even though it is the best estimator available, is likely to be very imprecise. Specifically, this means that other small samples obtained from the same population of patients will have substantially different measures of efficacy. Therefore, the only way that the investigators can have confidence in this estimate is if its very extreme.*

As we move to the right along the *x*-axis of Figure 5.10, information time increases. More data is accumulated, and the estimator begins to "settle down". The influence of sample-to-sample variability decreases as the number of clinical events increases. This increase improves the precision of the estimator, and the estimate of the treatment effect becomes more reliable. Thus, with more confidence in the estimator, the investigators can allow themselves to be persuaded by more moderate levels of efficacy than earlier in the study. The estimator is not likely to move wildly or unpredictably because its base is now on a larger number of clinical end-

* Examine Figure 4.3, page 87 to see how extreme effect size estimators can become in small samples.

points. The closer the trial is to its scheduled ending point, the more reliable is the data, and the less extreme its level of efficacy must be to carry persuasive weight with the investigators.

Investigators track the value of the monitoring device by tracking its position relative to the boundary on a graph (Figure 5.11).

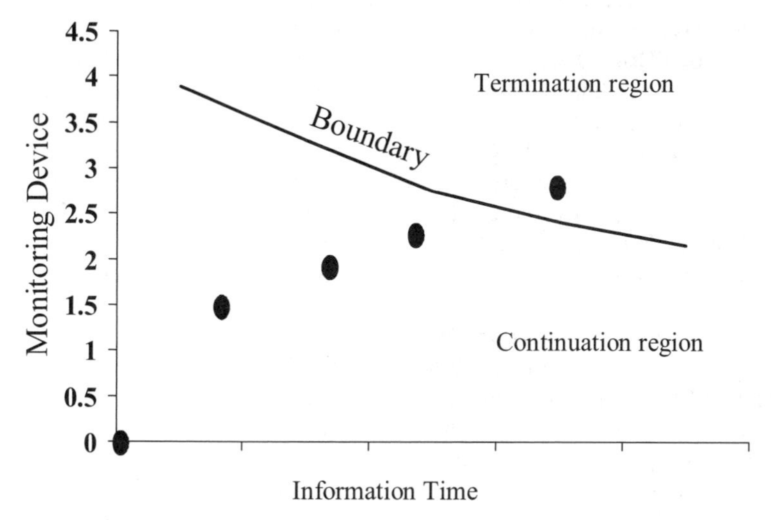

Figure 5.11. Successive values of the monitoring device over time in a clinical study. When the boundary is crossed, termination may be considered.

It was Canner [10] who suggested the idea of sloping boundary to decrease the likelihood that the monitoring device would exceed a boundary early in the trial. However, an important consideration in the computation of the boundary value is the type I error rate. How the investigators choose to allocate the type I error over the course of the study is a major determinant in the shape of the boundaries for early termination.

5.7.2 Alpha Spending Functions*

Both the motivation for and the shape of the boundary values for the monitoring device used to examine the interim results of clinical results have been elucidated at this point in our discussions. We will now turn our attention to how one actually computes the boundary values.

The need to conserve the overall type I error and the computing ability afforded by Brownian motion are central to this process. Recall from Chapter Four that the repeated testing of sample-based data actually inflates the level of the type I error rate, increasing the likelihood of drawing an erroneous conclusion from the study solely based on the vicissitudes of sampling error. We anticipate that there may be an opportunity to reduce this likelihood by taking into account a built-in dependency in the dataset. The source of this dependency lies in the group sequential process in which repeated analyzes are carried out on an accumulating collection of data. This data contains subjects that have been analyzed early in the study as well as new groups of data that have been recently collected and are being analyzed for the first time (Figure 5.1).

It was the paradigm-breaking research of O'Brien–Fleming that provided a relatively simple and easily understood procedure for converting a "single test clinical trial" (i.e., a trial that was designed to carry out a single hypothesis at its scheduled conclusion) into a monitored research program with alpha error conservation [11]. The O'Brien–Fleming chi-square test offered investigators the opportunity to monitor their clinical study on an ongoing basis, using Brownian motion considerations and the built-in dependency in group sequential data to control the overall type I error rate.* The Haybittle–Peto procedures [12,13] are also useful tools to develop interim procedures for clinical studies. However, they, like the O'Brien–Fleming procedures, require that the number of interim analyzes be prespecified. This work has been generalized by Wang and Tsiatis to demonstrate mathematical monitoring procedures that produce boundaries of a different variety of shapes [14].

What is important for the investigator to keep in mind is that they have an important voice in the shape of the boundary region. Mathematics in these fields can now support boundary computations to accommodate a wide range of considerations. The investigator, working with the statistician, should choose the shape of the boundary that will meet the ethical concerns that attend the interim monitoring procedures. Once this has been chosen, the statistician can then carry out the detailed computations.

The Lan–DeMets procedures [15] raised the idea of an alpha spending function to prominence. Simply, an *alpha spending function* is the relationship between the type I error rate and the information time at which that error is expended. Alpha spending functions distribute the type I error rate across the interim monitoring points of a clinical study. The determination of these type I error rates is made in a way that conserves the overall type I error.

The alpha spending functions are incorporated with Brownian motion concepts to construct the boundary values for monitoring the clinical trial. The smaller the type I error rate chosen for termination of a clinical trial at a particular boundary value, the greater the threshold must be for the test statistic for early termination consideration.

The modern use of the group sequential procedures has been nicely elaborated by Fleming [16]. The complicated mathematics are but one of a sequence of specific steps that investigators should follow in determining the group sequential procedures. The specific steps are

---

* When the boundary values generated by the O'Brien–Fleming group sequential procedures are graphed using the monitoring statistic as the y-axis, the boundary is flat. However, when the test statistic is used as the z-axis, the boundary decreases over time. See Jennison and Turnbull [9] for details.

Step 1: First select the overall type I error α_e.
Step 2: Set value of the interim monitoring alpha levels, $\alpha_1, \alpha_2, \alpha_3, \ldots, \alpha_k$ and so on.
Step 3: Choose critical regions for the monitoring device such that the probability that the first time the monitoring device falls in the critical region at interim monitoring point j is α_j.
Step 4: Monitor the research effort. Consider terminating the study at interim time point j if the monitoring device exceeds the boundary value at information time I_j.

Note that these steps in creating a boundary for the early conclusion of a study are most useful when executed in concordance with the three principles of sample-based research, elicited in Chapter Two.* Using a backward look procedure, the mathematics no longer require the investigator to pre-specify the number of interim monitoring points. However, concerns about the misleading affects of sampling error require the investigator and/or the DMC to state the number of a priori examinations there will be and to work diligently to keep the number of monitoring times to a minimum. This requires them to balance the effect of sampling error that crescendos as the number of interim examinations of the data increase on the one hand, with the ethical concern to examine the interim data as frequently as the ethics of the study require.

In the popular O'Brien–Fleming and Lan–DeMets paradigm in which the investigator asks "How likely is the path that test statistic has traversed from the beginning of the study under the null hypothesis?" these boundary values are available, but can be difficult to compute.

As a simple introductory example, consider an investigator who wishes to monitor a clinical program at information times t_1 and t_2 where $0 < t_1 < t_2 < 1$. She would like to compute boundary points for the monitoring device that, under the null hypothesis, follows standard Brownian motion in a way that conserves type I error. The investigator is interested in computing the boundary values b_1 and b_2 such that if $B(t_1) \geq b_1$ or $B(t_2) \geq b_2$, then there is sufficient evidence to consider terminating the trial for early efficacy. Extending to the scheduled end of the trial, we can also choose a value b_3 such that if $B(1) \geq b_3$, then the study ends with a positive result.

She proceeds under the assumption that she will end the study the first time the monitoring device $B(t)$ exceeds the boundary set for it. This means she would consider ending the study if $B(t_1) \geq b_1$. She would also end the study if $B(t_2) \geq b_2$ when at time information time t_1, $B(t_1)$ were less than b_1, Thus, she could set analysis specific alpha levels α_1, α_2, and α_3, such that

---

* Without following the principles of prospective declaration of a clear study design and analysis plan, the research results that are reported during an interim monitoring evaluation, regardless of effect size, revert to exploratory findings which must be repeated, commonly at great expense.

$$\begin{aligned} &\boldsymbol{P}\left[B(t_1) \ge b_1\right] = \alpha_1 \\ &\boldsymbol{P}\left[B(t_1) < b_1 \cap B(t_2) \ge b_2\right] = \alpha_2 \\ &\boldsymbol{P}\left[B(t_1) < b_1 \cap B(t_2) < b_2 \cap B(1) \ge b_3\right] = \alpha_3 \end{aligned} \tag{5.12}$$

Each of these events would be type I errors under the null hypothesis of no efficacy. Note that the use of joint probabilities allows us to incorporate the notion of dependence in our computations.

Clearly b_1 can be identified. Following the computation from Chapter Four, we can compute

$$\boldsymbol{P}\left[B(t_1) \ge b_1\right] = \boldsymbol{P}\left[N(0,t_1) \ge b_1\right] = \boldsymbol{P}\left[N(0,1) \ge \frac{b_1}{\sqrt{t_1}}\right] = \alpha_1.$$

Because we know from a consideration of the percentiles from a normal distribution that $\boldsymbol{P}\left[N(0,1) \ge Z_{1-\alpha_1}\right] = \alpha_1$, we have that $b_1 / \sqrt{t_1} = Z_{1-\alpha_1}$, or $b_1 = Z_{1-\alpha_1}\sqrt{t_1}$.

However, the second line of expression (5.12) introduces new complexities. It first requires us to identify the joint distribution of $B(t_1)$ and $B(t_2)$. Of course, $B(t_1)$ and $B(t_2)$ are not independent. The correlation coefficient reflecting their dependence is $\rho = \sqrt{t_1 / t_2}$.* Thus, we can write the mathematical function that can be used to compute joint probabilities involving $B(t_1)$ and $B(t_2)$. Specifically, if we let $x_1 = B(t_1)$ and $x_2 = B(t_2)$, then the formula used to compute probability that we seek is

$$f_{X_1,X_2}(x_1,x_2) = \frac{1}{2\pi\sqrt{t_1(t_2-t_1)}} e^{-\left(t_2 x_1^2 - 2t_1 x_1 x_2 + t_1 x_2^2\right) / 2t_1(t_2-t_1)}. \tag{5.13}$$

---

* The correlation between $B(t_1)$ and $B(t_2)$ can be identified by first finding the covariance of $x_1 = B(t_1)$ and $x_2 = B(t_2)$ and dividing by the product of the standard deviations of x_1 and x_2. Begin by writing $Cov(x_1,x_2) = E[x_1 x_2] - E[x_1]E[x_2] = E[x_1 x_2]$. The last equality holds because the mean of a standard Brownian motion process is zero. The joint expectation can be solved by invoking the double expectation argument that allows us to find means of functions of two or more variables, that is, $E\left[g(x_1 x_2)\right] = E_{x_1}\left[E_{x_2|x_1}\left[g(x_2 x_1)\right]\right]$. Applying this tool, we find $E[x_1 x_2] = E_{x_1}\left[E_{x_2|x_1}[x_2 x_1]\right] = E_{x_1}\left[x_1 E_{x_2|x_1}[x_2]\right] = E_{x_1}\left[x_1^2\right] = t_1$. The correlation follows

$$\rho = \frac{Cov(x_1,x_2)}{\sigma_{x_1}\sigma_{x_2}} = \frac{t_1}{\sqrt{t_1 t_2}} = \sqrt{\frac{t_1}{t_2}}.$$

Although the volume under the curve of this service is relatively easy to visualize (Figure 5.12), the evaluation of the volume under this surface, which measures the probability of the event of interest is complicated mathematically.*

In order to compute the values of the boundary, statisticians turn to computer packages that can carry out this quadrature, or numerical integration. These calculations directly approximate the areas (*r* volumes of regions) under these complicated curves. It is this style of calculation that is required to compute $\boldsymbol{P}\left[B(t_1) < b_1 \cap B(t_2) \geq b_2\right] = \alpha_2$ and to compute the more complicated probability $\boldsymbol{P}\left[B(t_1) < b_1 \cap B(t_2) < b_2 \cap B(1) \geq b_3\right] = \alpha_3$.

Example:
Open to the possibility that greater post-stroke physical activity may reduce the in-hospital recuperative following an acute stroke, investigators design a clinical trial to compare the three-week discharge rates of these patients. In the past, the proportion of patients who are hospitalized for more than three weeks is 11%. They hope to reduce this to 8.25%, that is, a 25% reduction.

The IRB requires the incorporation of an interim monitoring procedure into this study. The investigator plans to evaluate the interim data at five different

---

* The required computation is

$$\int_{\infty}^{b_1}\int_{b_2}^{\infty}\frac{1}{2\pi\sqrt{t_1(t_2-t_1)}}e^{-\left(t_2x_1^2-2t_1x_1x_2+t_1x_2^2\right)/2t_1(t_2-t_1)}\,dx_2dx_1.$$

For those who know calculus, this problem may be analytically approached in two ways. Unfortunately, neither approach yields an analytic solution. The first is to carry out the integration sequentially, by first integrating out the variable x_2. This process involves completing the square of the exponent, and thereby converting the integrand into the product of two functions of x_1. The first is a component of the normal distribution with respect to x_1. The second is the cumulative distribution function that is also a function of x_1. The result leads to the identification of the integral $\int_{-\infty}^{b_1} F_Z\left(g(x_1)\right)\frac{1}{\sqrt{2\pi v}}e^{-x_1^2/2v}\,dx_1$ where the functions g and the variance v are functions of t_1 and t_2. This can be integrated numerically but there is no closed form solution.

A second approach would be to write the variance-covariance matrix of x_1 and x_2 using standard matrix notation, that is $\boldsymbol{\Sigma} = \begin{bmatrix} t_1 & t_1 \\ t_1 & t_2 \end{bmatrix}$. This symmetric, positive definite matrix can be written in its principal form, that is, $\boldsymbol{\Sigma} = \mathbf{PDP'}$, where $\mathbf{P}$ is the matrix of orthonormal characteristic vectors of $\boldsymbol{\Sigma}$ and $\mathbf{D}$ is the diagonal matrix of characteristic values of $\boldsymbol{\Sigma}$. Using this approach, one can write $\boldsymbol{\Sigma} = \mathbf{PDP'} = \mathbf{PD}^{1/2}\mathbf{D}^{1/2}\mathbf{P'} = \boldsymbol{\Sigma}^{1/2}\boldsymbol{\Sigma}^{1/2'}$. This permits a linear transformation of the original correlated normally distributed random variables x_1 and x_2 into two uncorrelated independent variables w_1 and w_2 by $\underline{\mathbf{w}} = \begin{bmatrix} w_1 \\ w_2 \end{bmatrix} = \boldsymbol{\Sigma}^{-1/2'}\begin{bmatrix} x_1 \\ x_2 \end{bmatrix}$. The new region of integration is very complicated, and again, the integration does not yield a closed form solution.

information times (0.20, 0.40, 0.60, 0.80, 1.00). The investigator recognizes that the idea of terminating the study very early can only be justified with a high efficacy rate. Computations by Reboussin, Kim, DeMets and Lan[17]provide the boundaries (Table 5.1).

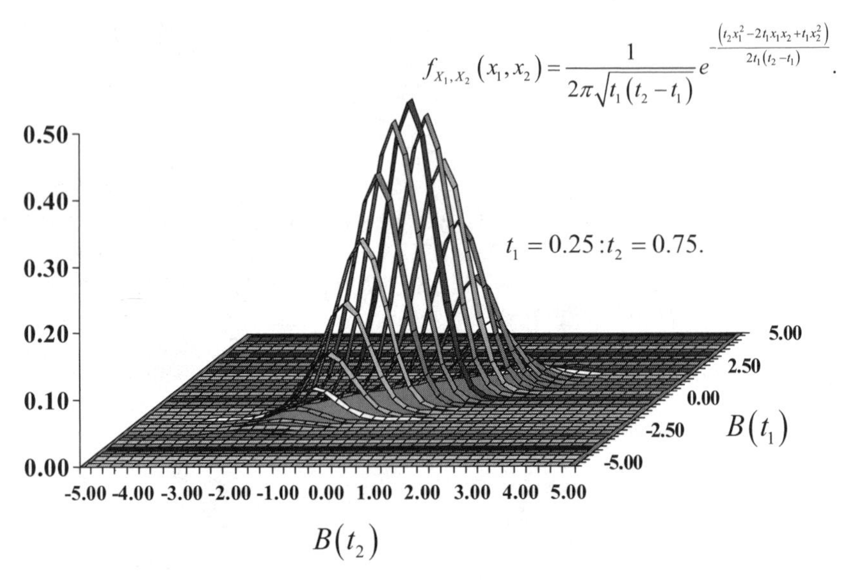

Figure 5.12. The surface whose enclosed volume for a specific region that must identified to find the boundary values in the look backward approach.

Table 5.1 Boundary Values for Test Statistic Comparing Mortality Rates in a Clinical Trial

Information Time	Boundary Value	Alpha
0.2	4.877	0.000001
0.4	3.357	0.000394
0.6	2.680	0.003681
0.8	2.290	0.011011
1.0	2.030	0.021178

Note the alpha spending function. Minimal alpha is used or "spent" at the early evaluation for $I = 0.20$. The overall type I error is controlled at the 0.05 level. Recall that, in this case, information time at point j is $I_j = {}^{n_j}\!/_N$, where n_j is the number of patients in the study at time point j and N is the total number of patients required to conclude the study. The graph of the test statistic versus time reveals the alpha spending function used for this evaluation (Figure 5.13). The investigator decides to review her data at information times 0.10, 0.50, 0.80. The boundary values for this process are easily depicted (Figure 5.13).

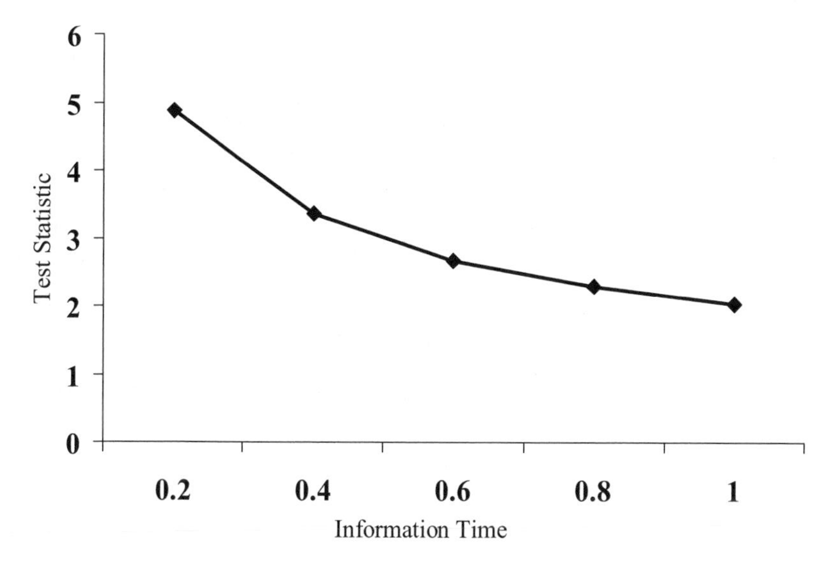

Figure 5.13. Efficacy boundary values for mortality rate in stroke trial.

Problems

1 Define the term group sequential procedure. Specifically how does the evaluation of data using a group sequential framework induce dependency in the analysis?

2. What are two major criticisms of the straight line or flat line boundary as a tool to use to monitor the efficacy of a study?

3. An investigator is interested in monitoring his study using an O'Brien–Fleming approach which is a Lan–DeMets type of alpha spending function. He will monitor the study at one and only one interim point. The investigators choose an interim value of the type I error α as $\alpha = 0.0005$. Compute and compare the value of the boundary of the Brownian monitoring procedure at each of $I = 0.10$, $I = 0.25$, and $I = 0.60$.

4. An investigator has put prospective interim monitoring rules in place for her clinical study. Under the alternative hypothesis, the anticipated drift in the monitoring device is $\mu = 0.15$. If the monitoring statistic has a value of 2.57 at information time $I = 0.10$, compute the likelihood of the null and alternative hypotheses. Repeat the computation for $I = 0.75$.

5. An investigator is interested in comparing the difference in left ventricular ejection fractions between Mexican-American (MA) and non-Hispanic White (NHW) patients in heart failure. Let x_i, i = 1, 2, 3, …, N represent the LVEF measurements on the MA patients, and let y_i, i = 1, 2, 3, …, N represent the LVEF measurements on the NHW subjects. Patients are recruited into the study, and have their ejection fractions calculated. It is anticipated that the study will require $2N$ patients, however, it will be monitored when $2n$ patients (n MA and n NHW subjects) have been observed in each group where $0 < n \leq N$. Let μ_{MA} and σ^2_{MA} be the population mean and variance of the LVEF for MA patients, and similarly, μ_{NHW} and σ^2_{NHW} be the corresponding parameters for NHW patients. Assume that both σ^2_{MA} and σ^2_{NHW} are known quantities. The clinical hypothesis test of interest is that the mean LVEF of MA is the same as that of NHW subjects.

   A. Following the development in Section 5.6.3, demonstrate that the monitoring statistic $B(I)$, where

$$B(I) = \frac{\sum_{i=1}^{n} x_i - \sum_{i=1}^{n} y_i}{\sqrt{N\left(\sigma^2_{MA} + \sigma^2_{NHW}\right)}}$$

   follows a normal distribution with variance $I = n/N$.

   B. Show that under the null hypothesis the mean value of $B(I)$ is zero.

   C. Demonstrate that, under the alternative hypothesis of $\mu_{MA} - \mu_{NHW} \neq 0$, the expected value of $B(I)$, $u_{B(I)}$ is

$$u_{B(I)} = \frac{\sqrt{N}\left(\mu_{MA} - \mu_{HW}\right)}{\sqrt{\sigma^2_{MA} + \sigma^2_{NHw}}} I.$$

   where $I = n/N$.

   D. Show that $TS(I) = \frac{B(I)}{\sqrt{I}}$ where

$$TS(I) = \frac{\overline{x_n} - \overline{y_n}}{\sqrt{\frac{1}{n}\left(\sigma^2_{MA} + \sigma^2_{NHW}\right)}}.$$

E. Compute the value of the monitoring statistic $B(I)$ when $n = 15$, $N = 150$, $\overline{X} = 57, \overline{Y} = 43$, $\sigma_{MA} = 15, \sigma_{NHW} = 20$. What is the information time for this calculation?

6. Assume that an investigator measures changes in carotid blood flow in a collection of N individuals over time. Let $x_{i,1,}$ be the first carotid blood flow measurement on the ith subject, and $x_{i,2}$ be the second carotid blood flow measurement. The investigator is interested in using these paired measurements to evaluate the change in carotid blood flow over time. Assume the difference in the mean carotid blood flow follows a normal distribution. Let the mean of the difference between carotid blood measures be μ_D and the variance of the difference be σ_D^2. The null hypothesis is H$_0$: $\mu_D = 0$ versus the alternative that the mean change in carotid blood flow is not zero.

A. Show that

$$B(I) = \frac{\sum_{i=1}^{n}\left(x_{i,1} - x_{i,2}\right)}{\sqrt{N}\sigma_D}$$

is normally distributed with mean $I\sqrt{N}\mu_D$ and variance I where $I = n/N$.

B. Show that $TS(I) = B(I)/\sqrt{I}$ where

$$TS(I) = \frac{\overline{x}_n - \overline{y}_n}{\sqrt{\frac{\sigma_D^2}{n}}}.$$

References

1. Pocock SJ. (1977) Group sequential methods in the design and analysis of clinical trials. *Biometrika.* **64**:191-199.
2. Wald A. (1947) *Sequential Analysis*. New York: John Wiley and Sons.
3. Armitage P. Sequential Medical Trials 2nd Edition (1975). London. Blackwell.
4. Slud E, Wei J. (1982) Two-sample repeated significance tests based on the modified Wilcoxon statistic. *Journal of the American Statistical Associations.* **77**:862-868.
5. Halperin M, Lan KKG., Ware JH, Johnson NJ, DeMets DL.(1982) An aid to data monitoring in long-term clinical trials. *Controlled Clinical Trials* **3**:311-323.
6. Sacks FM, Pfeffer MA, Moyé LA, Rouleau JL, Rutherford, JD, Cole TG, Brown L, Warnica JW, Arnold JMO, Wun CC, Davis BR, Braunwald E. for the Cholesterol and Recurrent Event Trial Investigators (1996). The effect of pravastatin on coronary events after myocardial infarction in patients with average cholesterol levels. *New England Journal of Medicine.* **335**:1001–1009.

7. Whitehead J. (1997) The Design and Analysis of Sequential Clinical Trials. Revised 2nd ed. Chicheser. Wiley.
8. Whitehead J, Stratton I. (1983) Group sequential clinical trials with triangular continuation regions. *Biometrics* **39**:227-236.
9. Jennison C, Turnbull BW. (1999) Group Sequential Procedures with Application to Clinical Trials. New York. Chapman Hall/CRC.
10. Canner PL. (1977) Monitoring Treatment Differences in Long-Term Cancer Trials. *Biometrics* **33**:603-615.
11. O'Brien PC., Fleming TR. (1979) A multiple testing procedure for clinical trials. *Biometrics* **35**:549–556.
12. Haybittle JL. (1971) Repeated assessment of results in clinical trials of cancer treatment *Br J Radiol* **44**: 793- 797.
13. Peto R, Pike MC, Armitage P, Breslow NE, Cox DR, Howard SV, Mantel N, McPherson K, Peto J, Smith PG. (1976) Design and analysis of randomized clinical trials requiring prolonged observation of each patient. I. Introduction and design. *Britisth Journal of Cancer* **34**: 585- 612.
14. Wang SK, and Tsiastis AA. (1987) Approximately optimal one-parameter boundaries for group sequential procedures. *Biometrics* **43**:193-200.
15. Lan KK, DeMets DL. (1983) Discrete sequential boundaries for clinical trials. *Biometrika* **70**: 659-663.
16. Fleming TR, Green SJ, Harrington DP. (1994) Considerations for Monitoring and Evaluating Treatment Effects in Clinical Trials. *Controlled Clinical Trials* **5**:55-66.
17. Reboussin DM, DeMets DL, Kim KM, and Lan KKG. Programs for computing group sequential boundaries using the Lan–DeMets methods. Version 2. Located at http://www.biostat.wisc.edu/landemets/simple.html.

6

Looking Forward: Conditional Power

Brownian motion is useful when monitoring clinical trials. It is naturally incorporated into the group sequential procedure approach, permitting us to take advantage of the indwelling dependence that we have seen is present in this set of calculations.

The focus that we have provided so far has been one where the investigator, interim monitoring statistic in hand, looks back over the course of the study completed thus far. This look-back assesses the likelihood of the path that the monitoring statistic has taken. The probability of this and similar paths can be computed under either the null hypothesis of no therapy effect, or an alternative hypothesis representing the presence of efficacy.

However, we have also observed that, except for the most elementary events, probability computations involving the look-back procedure can lead to complicated calculations requiring special software [1]. In addition, the computation of the boundary values built to prospectively guide efficacy determinations leads to computational complexity. Thus, although the traditional and highly valued look-back procedures have been and remain important tools in the modern monitoring of clinical research, the actual computations can be difficult for physician-investigators to carry out for themselves.

In most cases, this computational obstacle is of no real disadvantage to the clinical investigator. In the end, it is more important for the investigator to retain control over the event whose probability is computed than to focus on the underlying mathematics of the computation. However, there are circumstances in which the investigator is interested in computing the probability of several different events, and it can be most helpful if the investigator can actually carry out the computation, or at least, be able to see and understand how it works.

This chapter will focus on a collection of procedures that permit the investigator to compute probabilities and boundary values useful in monitoring ongoing clinical research, using a look-ahead procedure.

6.1 Notation and Perspective

We will continue to use the notation that we have developed. The trial will be monitored at a point I where $0 \le I \le 1$. The value of the test statistic at information time I is $TS(I)$. The value of the Brownian element is $B(I)$, where $B(I) = \sqrt{I}\,TS(I)$.

The look forward procedure focuses on predicting future paths of the test statistic using the Brownian motion framework. This is useful because, if some paths are extremely likely based on the information obtained thus far in the trial, the investigator may consider terminating the study early, precisely because of the high relative likelihood of this future trajectory (Figure 6.1).

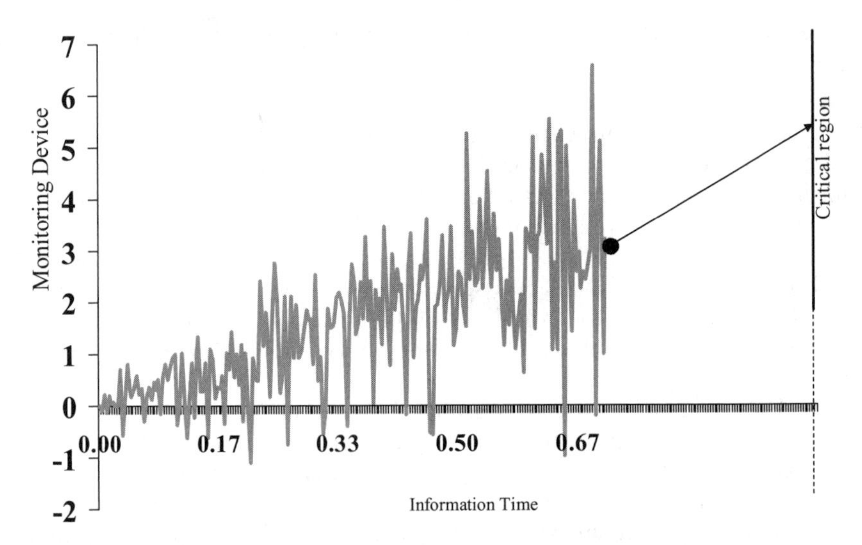

Figure 6.1. The conditional power approach uses assumptions about the trajectory of the monitoring device to project the likelihood of its future paths. This allows the investigator to compute the probability that the test statistic will fall in the critical region at the conclusion of the study, given the current value of the monitoring device.

From Figure 6.1, it is clear that there is a positive trajectory for the monitoring device. Regardless of whether the investigator assumes that the monitoring device continues its strong positive track into the future, or becomes a driftless process, simple computations involving Brownian motion permit the investigator to compute the probability of the test statistic landing in a region of interest at the study's end. If these computations suggest that the test statistic will fall in the critical region at the conclusion of the study, then consideration of termination procedures can begin.

These predictive computations can be carried out for any time point in the future. Of course, as we would expect, the farther into the future that we wish to project, the less accurate the projection will be.

6.2 Example of a "Forward Look" Computation

As an illustration, assume that an investigator in a clinical research effort has a test statistic of 2.30 at the $I = 0.50$ interim time point in the study. She would like to compute the probability that the test statistic will be at least as large as 1.96 at the 75% information time point. She begins the computation with the statement

$$\boldsymbol{P}\left[TS(0.75) \geq 1.96 \mid TS(0.50) = 2.30\right]. \tag{6.1}$$

The conditional probability in terms of the test statistic is difficult to solve, so we first convert this to Brownian motion, multiplying each event in expression (6.1) by $\sqrt{I}$. This reveals

$$\begin{aligned} &\boldsymbol{P}\left[\sqrt{0.75}\, TS(0.75) \geq \sqrt{0.75}\, 1.96 \mid \sqrt{0.50}\, TS(0.50) = \sqrt{0.50}\ 2.30\right] \\ &= \boldsymbol{P}\left[\sqrt{0.75}\, TS(0.75) \geq 1.697 \mid \sqrt{0.50}\, TS(0.50) = 1.626\right]. \end{aligned} \tag{6.2}$$

Because we know that $B(I) = \sqrt{I}\, TS(I)$ we may rewrite the last line of expression (6.2) as

$$\boldsymbol{P}\left[B(0.75) \geq 1.697 \mid B(0.50) = 1.626\right]. \tag{6.3}$$

Recalling our discussion in Chapter Four concerning Brownian motion conditioned on the past, expression (6.3) may now be written as

$$\boldsymbol{P}\left[B(0.75 - 0.50) \geq 1.697 - 1.626\right] = \boldsymbol{P}\left[B(0.25) = 0.071\right].$$

Under the null hypothesis of no treatment efficacy for the period of information time over which the prediction must be made,* that is, $0.50 \leq I \leq 0.75$, the calculation becomes

$$\begin{aligned} \boldsymbol{P}\left[B(0.25) \geq 0.071\right] &= \boldsymbol{P}\left[N(0, 0.25) \geq 0.071\right] \\ &= \boldsymbol{P}\left[N(0,1) \geq \frac{0.071}{\sqrt{0.25}}\right] = 1 - F_Z\left(0.142\right) = 0.444. \end{aligned}$$

Thus, the forward evaluation procedure suggests that there is a 44% chance that the test statistic will be greater than 1.96 when 75% of the information time has elapsed, given the value of the test statistic is 2.30 at $I = 0.50$.

The investigator can carry out this style of predictive computation for any future event of interest. For example, she may be interested in the value of the test

---

* We will have much more to say about this assumption later in this chapter.

statistic at the conclusion of the trial, based on its current value at $I = 0.50$. If she wishes to have the type I error for the final hypothesis test be 0.05, (i.e., the test statistic must be at least as large as 1.96 for a two-tailed evaluation), she may compute

$$\begin{aligned}
&\boldsymbol{P}\left[TS(1.00) \geq 1.96 \mid TS(0.50) = 2.30\right] \\
&= \boldsymbol{P}\left[\sqrt{1.00}\, TS(1.00) \geq \sqrt{1.00}\, 1.96 \mid \sqrt{0.50}\, TS(0.50) = \sqrt{0.50}\ 2.30\right] \\
&= \boldsymbol{P}\left[\sqrt{1.00}\, TS(1.00) \geq 1.96 \mid \sqrt{0.50}\, TS(0.50) = 1.626\right] \\
&= \boldsymbol{P}\left[B(1.00) \geq 1.96 \mid B(0.50) = 1.626\right] \qquad (6.4) \\
&= \boldsymbol{P}\left[B(1.00 - 0.50) \geq 1.96 - 1.626\right] = \boldsymbol{P}\left[B(0.50) = 0.334\right] \\
&= \boldsymbol{P}\left[B(0.50) \geq 0.334\right] = \boldsymbol{P}\left[N(0, 0.50)\ \geq\ 0.334\right] \\
&= \boldsymbol{P}\left[N(0,1) \geq \frac{0.334}{\sqrt{0.50}}\right] = 1 - F_Z\left(0.472\right) = 0.318,
\end{aligned}$$

which is the probability that the test statistic is in the critical region at the end of the study. Comparing this answer to the $\boldsymbol{P}\left[TS(0.75) \geq 1.96 \mid TS(0.50) = 2.30\right] = 0.444$, we observe that the probability that the value of the test statistic at $I = 0.75$ will be greater than 1.96 given its value of 2.30 is actually greater than the analogous $\boldsymbol{P}\left[TS(1.00) \geq 1.96 \mid TS(0.50) = 2.30\right]$.

This bears some examination. The region that the test statistics will fall in for each of these computations is the same. What is different is the period of time over which the prediction is required. Specifically, the farther into the future one wishes to predict, the more variability is associated with the location of the test statistic.

6.3 Nomenclature

This type of forward look procedure is known among statisticians and clinical trial specialists as *conditional power*. As is the case with many of the interim monitoring procedures, there are many contributors to the development of this approach.

The philosophy first appeared in industry as quality control specialists turned their attention to screening a lot of n manufactured items for acceptability. The decision rule used at the time claimed that the lot was acceptable if it contained less than c defectives; that is, brand the lot as unacceptable if it contained c or more defective items. However, the lot must be acceptable if there are $n - (c - 1)$ non-defects,* and so it was suggested that by monitoring the number of nondefectives, the sample of tested items could be reduced. This reduction is described as curtailment [2]. One could curtail the examination by scanning for non-defectives.

---

* The lot meets the bare minimum requirement of acceptability if there are c defective items, and $n - c$ non-defective items. If there are $n - c + 1$ nondefectives, then the lot must be acceptable.

In this industrial circumstance, the scanning ceased as soon as the result became inevitable. However, why wait for inevitability? Interim probability computations would allow one to compute how likely it is that the lot would be acceptable based on the results of the lot's partial scan. This modification was termed *stochastic curtailment.* It is this expression that was transferred to the clinical trial arena and is now used as a moniker for the look forward procedure. A somewhat more user-friendly descriptor has come to be known as *conditional power* [3].*

The conditional power computation that we have illustrated thus far can be generalized. Assume the value of the test statistic at information time I_1 is s_1, that is, $TS(I_1) = s_1$. We are interesting in the probability that $TS(I_2) \geq s_2$ for $0 \leq I_1 \leq I_2 \leq 1$, or

$$\boldsymbol{P}\left[TS(I_2) \geq s_2 \mid TS(I_1) = s_1\right]. \tag{6.5}$$

We first convert this event to an event involving the monitoring device that follows standard Brownian motion under the null hypothesis.

$$\begin{aligned}\boldsymbol{P}\left[TS(I_2) \geq s_2 \mid TS(I_1) = s_1\right] &= \boldsymbol{P}\left[\sqrt{I_2}TS(I_2) \geq \sqrt{I_2}s_2 \mid \sqrt{I_1}TS(I_1) = \sqrt{I_1}s_1\right] \\ &= \boldsymbol{P}\left[B(I_2) \geq \sqrt{I_2}s_2 \mid B(I_1) = \sqrt{I_1}s_1\right].\end{aligned}$$

We now use what we know about Brownian motion conditioned on the past to write

$$\begin{aligned}\boldsymbol{P}\left[B(I_2) \geq \sqrt{I_2}s_2 \mid B(I_1) = \sqrt{I_1}s_1\right] &= \boldsymbol{P}\left[B(I_2 - I_1) \geq \sqrt{I_2}s_2 - \sqrt{I_1}s_1\right] \\ &= \boldsymbol{P}\left[N(0, I_2 - I_1) \geq \sqrt{I_2}s_2 - \sqrt{I_1}s_1\right] \\ &= \boldsymbol{P}\left[N(0,1) \geq \frac{\sqrt{I_2}s_2 - \sqrt{I_1}s_1}{\sqrt{I_2 - I_1}}\right] \\ &= 1 - F_Z\left[\frac{\sqrt{I_2}s_2 - \sqrt{I_1}s_1}{\sqrt{I_2 - I_1}}\right].\end{aligned} \tag{6.6}$$

When $I_2 = 1$, this simplifies to

$$1 - F_Z\left[\frac{s_2 - \sqrt{I_1}s_1}{\sqrt{1 - I_1}}\right]. \tag{6.7}$$

Recall that this calculation is carried out under the hypothesis of standard Brownian motion, that is, under the null hypothesis of no treatment effect. Thus, the standard Brownian motion assumption in the conditional power computation as-

---

* Bayesians have re-worked this notion, describing their adaptation as *predictive power*. This is described in Chapter Eight.

sumes that any research effect in the study has already been seen. This is known as conditional power computed under the null hypothesis, or $C_P(H_0)$.

Conditional power may also be computed under the alternative hypothesis. Under this alternative hypothesis, we assume that the Brownian element experiences drift with parameter μ. In this case

$$\begin{aligned}
&\boldsymbol{P}\left[TS(I_2)\geq s_2 \mid TS(I_1)=s_1\right]\\
&=\boldsymbol{P}\left[TS(I_2)\geq s_2 \mid TS(I_1)=s_1\right]=\boldsymbol{P}\left[\sqrt{I_2}TS(I_2)\geq \sqrt{I_2}s_2 \mid \sqrt{I_1}TS(I_1)=\sqrt{I_1}s_1\right]\\
&=\boldsymbol{P}\left[B(I_2)\geq \sqrt{I_2}s_2 \mid B(I_1)=\sqrt{I_1}s_1\right].\\
&=\boldsymbol{P}\left[B(I_2)\geq \sqrt{I_2}s_2 \mid B(I_1)=\sqrt{I_1}s_1\right]=\boldsymbol{P}\left[B(I_2-I_1)\geq \sqrt{I_2}s_2-\sqrt{I_1}s_1\right].
\end{aligned}$$

As this point, the drift parameter must be incorporated.

$$\begin{aligned}
\boldsymbol{P}\left[B(I_2-I_1)\geq \sqrt{I_2}s_2-\sqrt{I_1}s_1\right]&=\boldsymbol{P}\left[N\left(\mu(I_2-I_1),I_2-I_1\right)\geq \sqrt{I_2}s_2-\sqrt{I_1}s_1\right]\\
&=\boldsymbol{P}\left[N(0,1)\geq \frac{\sqrt{I_2}s_2-\sqrt{I_1}s_1-\mu(I_2-I_1)}{\sqrt{I_2-I_1}}\right] \qquad (6.8)\\
&=1-F_Z\left[\frac{\sqrt{I_2}s_2-\sqrt{I_1}s_1-\mu(I_2-I_1)}{\sqrt{I_2-I_1}}\right].
\end{aligned}$$

When I_2 =1, and we wish to predict to the end of the study, we find

$$C_P(H_a)=1-F_Z\left[\frac{s_2-\sqrt{I_1}s_1-\mu(1-I_1)}{\sqrt{1-I_1}}\right]. \qquad (6.9)$$

This style of computation was popularized by Lan and Wittes [4].

6.4. Examples of Conditional Power Computations

As an example, consider an investigator who is interested in computing conditional power for a clinical trial designed to assess the effect of therapy on the recurrent fatal and nonfatal stroke rate of patients who have suffered a stroke. In this study, patients who have had a stroke within the past six months are randomly allocated to receive either active or control group therapy. They are then followed for three years, during which time the number of fatal and nonfatal strokes is compiled. At the conclusion of the study, the investigators plan to carry out a standard survival analysis, comparing the recurrent stroke rate in the active group with the stroke rate in the placebo group.

It is anticipated that the incidence of fatal and nonfatal strokes in the control group will be 23%. The investigators anticipate that there were be a 20% reduction in this rate in the active group, resulting in a fatal/nonfatal stroke rate of $(0.23)(1-.20) = 0.184 = 18.4\%$ in the active group. This leads to sample size (number of patients in the active group + number of patients in the control group) of 2778.

The investigators plan to execute a standard lifetable analysis of the data, intending to calculate a log rank test statistic at the conclusion of the study. However, the study's DMC will monitor this log rank statistic over the course of the study, using a conditional power computation to monitor for the occurrence of unanticipated early efficacy. The study is to be conducted with a type I error rate of 0.04 at the conclusion of the study, requiring a test statistic at the conclusion of the study of at least 2.054.*

In a trial where the endpoint is dichotomous, the information time is measured as the proportion of endpoint events that have occurred in the study [5]. For this example, information time is proportional to the number of fatal/nonfatal strokes that have occurred. The total number of strokes that are expected is the total number anticipated in the active group $\left(\left(2778/2\right)(0.184) = 256\right)$ plus the number of fatal and nonfatal strokes that are anticipated to occur in patients recruited to the control group $\left(\left(2778/2\right)(0.23) = 282\right)$ or 538.

At an interim monitoring point in the study, the DMC is provided with the following information. There have been 424 fatal/nonfatal strokes in the study, and the test statistic is 2.5. They would like to compute the conditional power of this finding under a variety of assumptions.

Their first computation is under the assumption that no further effectiveness will be seen for the remainder of the trial. The information time for this interim examination is $424/538 = 0.788$ or 78.8%. This is $C_P(H_0)$ and is computed using formula (6.7) as

$$\begin{aligned} \boldsymbol{P}\left[TS(1) \geq 2.054 \mid TS(0.788) = 2.5\right] &= 1 - F_Z\left[\frac{2.054 - \sqrt{0.788}\,(2.5)}{\sqrt{1-0.788}}\right] \\ &= 1 - F_Z\left[-0.358\right] = 0.642. \end{aligned} \tag{6.10}$$

There is a 64.2% chance that the log rank test statistic will fall in the critical region at the conclusion of the study, given that its current value at 78.8% information time is 2.5. However, this computation is based on the assumption that no further efficacy is experienced for the duration of the time. This means the events will occur in the active and control groups at the same rate.

In order to explore additional scenarios, the DMC computes the conditional power under the assumption that the relative risk for the occurrence of fatal/nonfatal strokes for the duration of the study is 0.80, corresponding to a 20%

* The issue of type I error conservation under conditional power will be discussed in the next section.

reduction of the fatal/nonfatal stroke rate, consistent with the alternative hypothesis. The conditional power computation must now proceed under the alternative hypothesis according to (6.9). The drift parameter μ for this study is computed from the relative risk for the remainder of the trial R, and the total number of events (in this case the total number of fatal and nonfatal strokes in the study) (E) as

$$\mu = -\ln(R)\sqrt{E/4}. \tag{6.11}$$

Using this formulation produces the following conditional power computation.

$$\begin{aligned}
CP(H_a) &= 1 - F_Z\left[\frac{s_2 - \sqrt{I_1}s_1 - \mu(1-I_1)}{\sqrt{1-I_1}}\right] \\
&= 1 - F_Z\left[\frac{2.054 - \sqrt{0.788}(2.5) - \left(-\ln(0.80)\sqrt{\frac{538}{4}}(1-0.788)\right)}{\sqrt{1-0.788}}\right] \\
&= 1 - F_Z[-1.55] \\
&= 0.939.
\end{aligned}$$

This finding can produce vigorous debate for the DMC. Under the alternative hypothesis, there is substantial conditional power, and some might argue that early termination of the study should be implemented. Alternatively, a conservative stance would be to accept the computation under the null hypothesis; since there is only 64.2% conditional power in this scenario, the study result is not inevitable, and the trial should be continued. In fact, conditional power can be computed under a range of alternative hypotheses of the relative risk (Figure 6.2).

The best advice in this circumstance is for the DMC to integrate these conditional power computations with other information available to it before it makes a recommendation.. For example, the announcement of the findings of a companion study demonstrating the effectiveness of the same intervention in a similar but independent sample of patients may improve the persuasiveness of the lower $C_P(H_o)$ calculation. Alternatively, the appearance of an important adverse effect associated with the intervention being tested may convince the DMC that there must be ironclad assurance that the intervention's benefits will adumbrate its risks. In this case, the DMC may argue that insufficient conditional power has been demonstrated by the $C_P(H_0)$ computation to justify early termination, and that the study should continue.

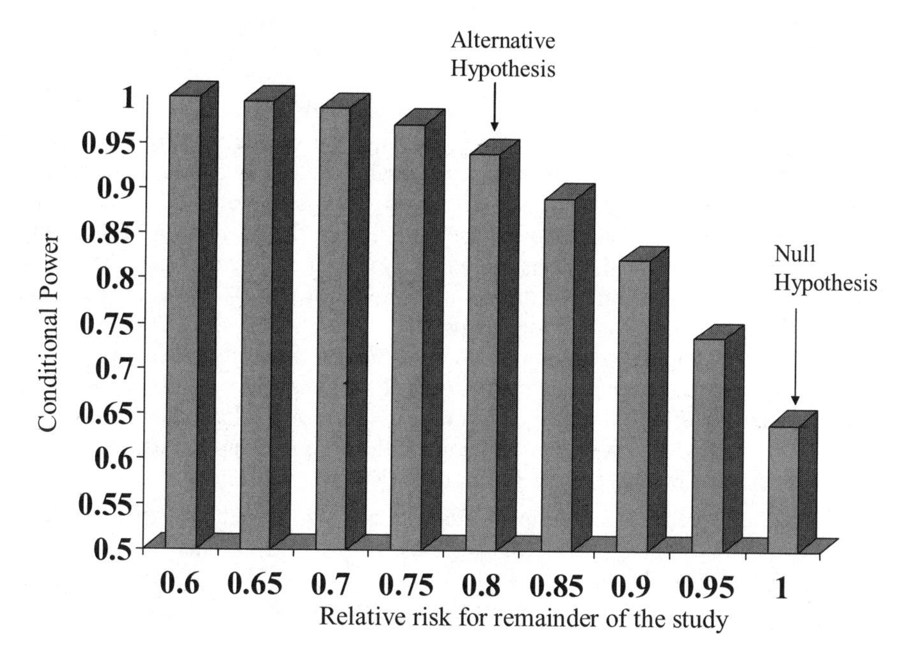

Figure 6.2. Conditional power as a function of the relative risk for the remainder of the trial (overall alpha = 0.04, 78.8% information time).

In each circumstance the statistical monitoring computation is just the beginning and not the end of the deliberation for early termination of the trial.

6.5 Type I Error Under Conditional Power

One of the attractive features of the conditional power calculation (in fact, the same attractive feature that is common to all group sequential procedures) is its ability to directly compute the probability of complicated events that are of interest to investigators. However, being cognizant of the need for overall type I error rate control requires that our attention turn to the issue of multiple comparisons.

Let α be the type I error rate that is set for the hypothesis test at the conclusion of the trial. Let ξ be the overall type I error rate, that is, the probability of making at least one type I error from the k interim evaluations that are performed during the course of the study. We know that, in general $\xi \geq \alpha$ due to the fact that multiple tests are carried out.* However, the built-in dependence in the Brownian motion process produces substantial saving for the total type I error expended during the course of the interim evaluations. If γ is the conditional power of the study, then Grimmett and Stirzaker [6] demonstrated that the overall type I error rate of the study ξ is related to α and γ by

* Recall that, in the independence assumption in which k hypothesis tests are carried out, ξ can be approximated by $\xi = k\alpha$ for small α.

$$\xi \leq \frac{\alpha}{\gamma}. \tag{6.12}$$

One important implication of this finding relates to the magnitude of the type I error rate penalty one must pay for interim evaluations. In environments where the conditional power is large, the overall penalty that one pays for the use of this interim monitoring procedure is relatively small. For example, if a researcher is interested in carrying out a single hypothesis test at the conclusion of the study at the $\xi = 0.05$ level, but she plans to consider terminating the trial early if $C_P(H_0) = 0.95$, then the maximum total type I alpha expended is 0.052. This represents a small inflation in the overall type I error rate.

This simple calculation is more useful if it is implemented prospectively. In this circumstance, the investigator would first choose the overall type I error rate for the research effort, and the conditional power level at which she would consider terminating the study. With these two quantities, she can approximate the test specific alpha error rate at the conclusion of the study by $\alpha \leq \xi\gamma$. For example, if she wishes to control the overall alpha error rate at 0.05, then the type I error rate for the hypothesis test being monitored is (0.05)(0.95) = 0.0475. This value is then used as the alpha threshold on which the individual computations for conditional power should be based. Thus, from expression (6.7), she computes $C_P(H_0)$ as

$$\begin{aligned} C_P(H_0) &= \boldsymbol{P}\left[TS(1) \geq s_2 \mid TS(I_1) = s_1\right] \\ &= \boldsymbol{P}\left[TS(1) \geq Z_{1-0.0475/2} \mid TS(I_1) = s_1\right] \\ &= 1 - F_Z\left[\frac{1.982 - \sqrt{I_1}\, s_1}{\sqrt{1-I_1}}\right]. \end{aligned}$$

Note that the test statistic of 1.982 is not very different from the value of 1.96 that would be the traditional value for the Z percentile if no correction for multiplicity were required. If the value of the test statistic at $I_1 = 0.65$ is 2.8, then

$$\begin{aligned} C_P(H_0) = 1 - F_Z\left[\frac{1.982 - \sqrt{I_1}\, s_1}{\sqrt{1-I_1}}\right] &= 1 - F_Z\left[\frac{1.982 - \sqrt{0.65}\; 2.80}{\sqrt{1-0.65}}\right] \\ = 1 - F_Z\left[\frac{1.982 - \sqrt{0.65}\; 2.80}{\sqrt{1-0.65}}\right] &= 1 - F_Z\left[-0.466\right] = 0.679. \end{aligned}$$

A second implication of the computation $\xi \leq \alpha/\gamma$ is that this multiplicity correction for conditional power is not a function of the number of interim evaluations of the trial. Whether the investigations evaluate the data at one interim point or evaluate the data repeatedly during the course of the trial, the multiplicity correction remains the same.

However, seasoned investigators understand the dangers of interpreting this observation as carte blanche authority to evaluate the interim data in a frequent and unplanned fashion during the trial's conduct. Although ξ is well controlled for multiple interim data evaluations, sampling error can still provide misleading information about the magnitude of efficacy, as well as the incidence of adverse events. Thus, even in the face of the relatively small correction for multiplicity of type I error in the conditional power environment, a prospectively planned, disciplined, and well-structured monitoring process reduces the likelihood of erroneous, sample-error-driven results.

6.6 Boundary Values and Conditional Power

Boundary values identified by investigators that provide guidance in monitoring clinical studies are readily available from conditional power analyzes. In this circumstance, the investigators wish to compute the value b such that the probability that the test statistic falls in the critical region at the conclusion of the study (when the test statistic equals b at information time I_1) is at least some value γ (i.e., γ = 95%) . The probability statement for this occurrence is

$$\boldsymbol{P}\left[TS(1) \geq Z_{1-\alpha/2} \mid TS(I_1) = b\right] = \gamma. \tag{6.13}$$

The solution of this problem is straightforward and appears in Appendix C.

$$b = \frac{Z_{1-\alpha/2} - Z_{1-\gamma}\sqrt{1-I_1}}{\sqrt{I_1}}. \tag{6.14}$$

The easily implemented equation (6.14) produces readily computable boundary values. An examination of the value of these boundaries as a function of both information time and conditional power reveals some expected features (Figure 6.3).

We can observe two important characteristics of the curves presented in Figure 6.3. The first is that, for a given conditional power, the boundary values decrease as a function of information time. This is consistent with the pattern we have seen for other group sequential procedures, and is reflective of the relative imprecision of estimators obtained early in a clinical research effort. This imprecision leads us to reject modest levels of efficacy that may appear initially in a research effort as sufficient reason to discontinue a study.

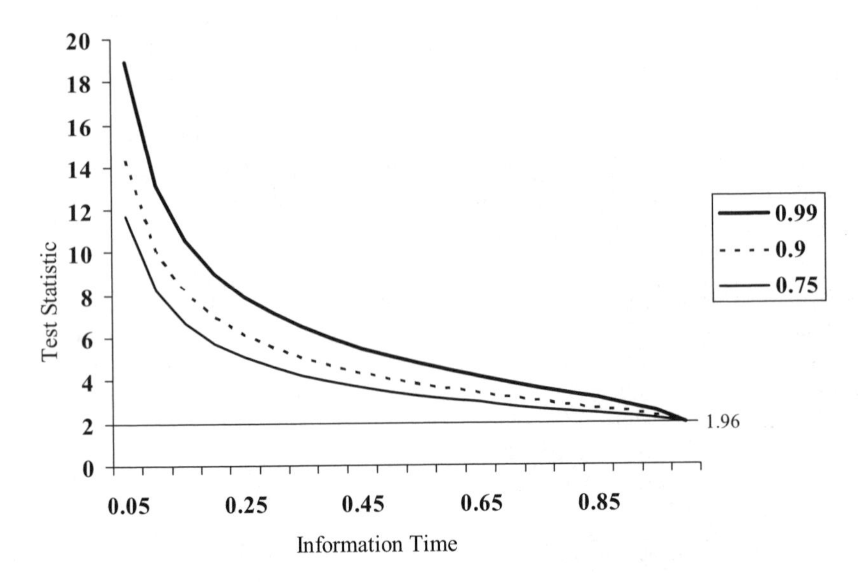

Figure 6.3. Boundary values as a function of conditional power (legend) and information time. Type I error = 0.05 (two-tailed).

Secondly, Figure 6.3 demonstrates the relationship between the magnitude of the boundary value and the level of conditional power. Essentially, the greater the conditional power, the greater the efficacy level required to cross the boundary for consideration of early termination. This observation is actually a natural consequence of the definition of conditional power. The larger the conditional power value, the greater the likelihood of a positive conclusion at the end of the study based on the value of the interim result. Essentially, the more "inevitable" the investigator wishes a positive research conclusion to be, the stronger must be the measure of efficacy of the current data.

6.7 Conditional Power Computations

As an example of the use of these boundary values, consider a group of investigators interested in examining the effect of a new therapy that will reduce the occurrence of fatal and nonfatal strokes. This will be a randomized, controlled clinical trial. In the population that is being studied, the cumulative four-year fatal/nonfatal stroke rate is expected to be 18%. The investigators anticipate that the intervention will reduce this rate by 25%. Assuming a two-sided type I error rate of 0.05, and a power of 90%, the required sample size of the study is 2745 patients.

After these preliminary computations, the investigators turn to incorporation of conditional power-based interim monitoring procedures. They are willing to consider stopping the trial for early benefit* if their interim results provide at least 90% conditional power, that is, the probability that the test statistic will fall in the critical region at the conclusion of the study (based on the interim result) is at least 90%.

An early first step to incorporate conditional power is to adjust the computation of the familywise error for the multiple evaluations that will take place during the course of the study. Using formula (6.12), the investigators first adjust the overall familywise error rate down from $\xi = 0.05$ to $\xi = (0.05)(0.90) = 0.045$. Thus, the evaluation at the end of study must be carried out at the 0.045 level. This adjusted type I error rate produces a new sample size computation of 2822 subjects required for the study, a change that represents a $(2822 - 2745)/2745 = 2.8\%$ increase.

Knowledge of the total number of patients who experience at least one event in the study is necessary in order to compute information time accrual. The total number of patients with events in the control group is $\left(2822/2\right)(0.18) = 254.$ The total number of events in the active group is $\left(2822/2\right)(0.75)(0.18) = 191$, resulting in 254 + 191 = 445 total fatal/nonfatal strokes required in the study. This will be used as the denominator of the information time calculations that the investigators will carry out.

The investigators decide to provide interim evaluations at 25%, 50%, and 90% information time, corresponding to the points when 112, 223, and 400 events have accrued, respectively. The overall type I error rate, conditional power, and information time are used to compute the boundary values using equation (6.14) (Table 6.1).

Table 6.1. Upper Boundary Values for Early Termination (Two Sded, Type I Error Rate = 0.045, 90% Conditional Power)

Information Time	Number of Events	Upper Boundary
0.25	111	6.23
0.50	223	4.12
0.90	401	2.54

During the study, at the 50% information time point, the investigators observe that the test statistic is 4.54. Because this is greater than the boundary value of 4.12, they have met the statistical criteria for discontinuing the study. However, as we have seen, the investigators would be wise to consider the entire weight of evidence before they decide to stop the study. This includes, but is not limited to: (1) the magnitude of the effect size (in this case, the risk reduction in the occurrence of fatal and nonfatal strokes, (2) the effect of the intervention on secondary endpoints, and (3) the occurrence of adverse events reasonably believed to be associated with

* Consideration for stopping for harm is covered in Chapter Seven.

the intervention. The joint consideration of these events would lead to a more fully informed decision to terminate the study.

6.7.1 Monitoring the Number of Events*

Occasionally, an investigator is interested in computing a boundary value that is not based on the test statistics, but on the number of events in a clinical study. For example, a researcher is carrying out an assessment of the rate of intracerebral hemorrhages in patients treated for acute stroke. She believes the underlying rate is p_0. At the conclusion of her work, the investigator will have recruited and followed N patients. However, she would like to monitor the study at interim information time $I = n/N$ when n patients have been observed. We now know that we can compute a conditional power-based boundary value. Specifically, if she is interested in the conditional power $\boldsymbol{P}\left[TS(1) \geq Z_{1-\alpha/2} \mid TS(I_1) = b\right] = \gamma$, then the boundary value for this procedure is

$$b = \frac{Z_{1-\alpha/2} - Z_{1-\gamma}\sqrt{1-I_1}}{\sqrt{I_1}}. \tag{6.15}$$

This is a boundary value in terms of the test statistic. However, the investigators are interested not in a boundary value for the test statistic, but in the boundary value for the number of intracerebral hemorrhages. Because the test statistics at information time $TS(I)$ based on n observations is

$$\frac{\sum_{i=1}^{n} x_i - np_0}{\sqrt{np_0(1-p_0)}}$$

we may write

$$\frac{\sum_{i=1}^{n} x_i - np_0}{\sqrt{np_0(1-p_0)}} = b$$

or

$$\sum_{i=1}^{n} x_i = np_0 + b\sqrt{np_0(1-p_0)}.$$

Incorporating the value of b we find that the number of intracerebral hemorrhages is

$$\sum_{i=1}^{n} x_i = np_0 + \frac{Z_{1-\alpha/2} - Z_{1-\gamma}\sqrt{1-I}}{\sqrt{I}}\sqrt{np_0(1-p_0)}.$$

6.8 Following Trajectories: Brownian Bridges*

During the interim monitoring of clinical research, investigators commonly have more questions about the trajectory of their research results than whether the results will fall in the critical region. They may be concerned about whether a particular trajectory continues on what may be a path to a value that would trigger some action before the study is over.

For example, consider a randomized clinical trial that is designed to evaluate the effect of therapy on the combined endpoint rates of total mortality and total hospitalizations. The investigators hope to reduce the cumulative incidence of this combined endpoint by 20% from its one-year incidence of 15%. The sample size for the trial, assuming a power of 90%, conditional power of 95%, and a two-sided type I error rate of 0.05 (adjusted for conditional power to 0.048) is 5520. They plan to evaluate the study results at 20%, 50%, 75%, and 90% information times. Using equation (6.14), they compute the following boundary values for efficacy (Table 6.2).

Table 6.2. Upper Boundary Values for Early Termination (Two Sded, Type I Error Rate = 0.045, 95% Conditional Power)

Information Time	Number of Events	Upper Boundary
0.25	186	6.81
0.50	373	4.45
0.75	559	3.24
0.90	671	2.64

The trial proceeds as planned, and the investigators observe that $TS(0.25) = 2.8$. In this situation (Scenario 1), they are interested in whether the boundary value of will be reached at $I_2 = 0.50$.

A simple conditional power computation produces the result. From equation (6.6), this is readily computed to be

$$\begin{aligned} C_P(H_0) &= \boldsymbol{P}\left[TS(0.50) > 4.45 \mid TS(0.25) = 2.8\right] \\ &= 1 - F_Z\left[\frac{\sqrt{0.50}\,(4.45) - \sqrt{0.25}\,(2.8)}{\sqrt{0.50 - 0.25}}\right] \\ &= 1 - F_Z(3.49) = 0. \end{aligned}$$

The low value of this probability should come as no surprise; it is computed conservatively that is, it assumes that there is no further demonstration of efficacy after the $I_1 = 0.25$ information time point. An alternative computation in this setting is $C_P(H_a)$. In fact, the conditional power computation under each assumption is useful and informative (Figure 6.4).

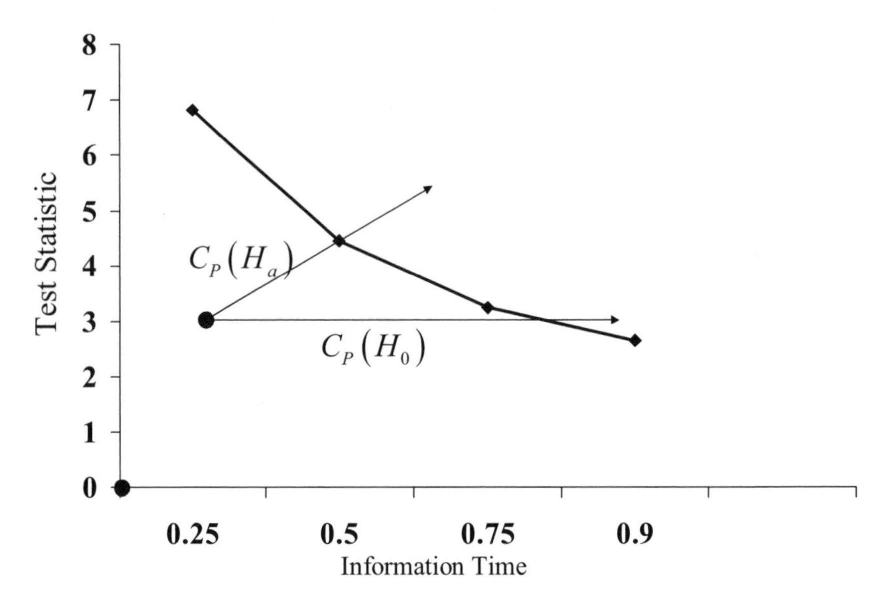

Figure 6.4. Mapping the trajectory of the test statistic relative to the conditional power boundary values under the null and alternative hypotheses.

For this computation, the drift parameter is required. Recall from expression (6.11) that the drift parameter $\mu = -\ln(RR)\sqrt{E/4}$. In this example, the relative risk of 0.80 (corresponding to a 20% reduction in events) and $E = 745$ produces a drift parameter of 3.045. Thus, $C_P(H_a)$ may be computed from equation (6.8) as

$$\begin{aligned} C_P(H_a) &= \boldsymbol{P}\left[TS(0.50) > 4.45 \mid TS(0.25) = 2.8\right] \\ &= 1 - F_Z\left[\frac{\sqrt{I_2}s_2 - \sqrt{I_1}s_1 - \mu(I_2 - I_1)}{\sqrt{1 - I_1}}\right] \\ &= 1 - F_Z\left[\frac{\sqrt{0.50}(4.45) - \sqrt{0.25}(2.8) - 3.045(0.50)}{\sqrt{0.50 - 0.25}}\right] \\ &= 1 - F_Z(0.448) = 0.327. \end{aligned}$$

Under the alternative hypothesis, the probability of crossing the boundary at $I = 0.50$ is 0.327. While still not likely, this probability is substantially greater than the probability of crossing the boundary under the null hypothesis.

Alternatively, the scientists monitoring the trial may be interested in tracking the test statistic's path after it has already crossed a monitoring boundary (Scenario 2, Figure 6.5).

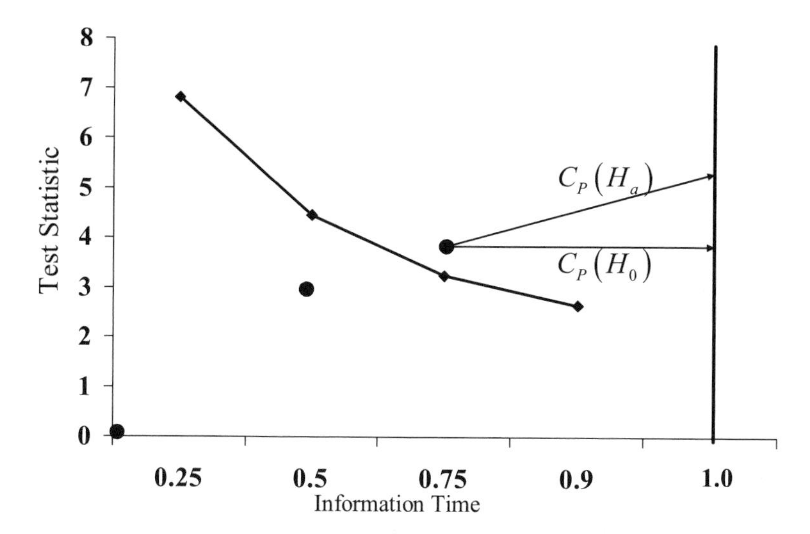

Figure 6.5. Mapping the future trajectory of the test statistic relative to the end of the study under the null and alternative hypotheses.

In this latter circumstance, even though the monitoring boundary was crossed, the investigators nevertheless continued the study. This might be justified in order to (1) collect information on secondary endpoints, (2) to precisely measure the uniformity of treatment effects on subgroups, or (3) to gather additional safety information.

These computations provide the probability that the movement of a Brownian element that is fixed at one value (the current value of the transformed test statistic), will fall in a region at a particular information time point in the future. The computation makes no assumption about the occurrence of monitoring between information time points 0.25 and 0.50; they produce only an estimate of the probability that the value of the test statistic will exceed a value at information time 0.50. In fact, no assumption is made about the value of the test statistic at any information time point I lying between 0.25 and 0.50.

An alternative would be to map the anticipated path that the investigators believe the test statistic will follow. Specifically, there may be specific interest in determining if the test statistic will remain on its current trajectory [7]. This is an activity that might be most easily executed by those whose task it is to produce and monitor the test statistic on a more frequent basis than the DMC (e.g., the coordinating center or data center for the study).

This type of computation is produced from what is known as *tied-down Brownian motion*. Tied-down Brownian motion, or a *Brownian bridge* is simply a generalization of Brownian motion that was conditioned on the past. Recall that, if $B(t)$ is standard Brownian motion and its value at time t_1 is known where $t_1 \leq t$, then the probability distribution of $B(t)$ conditioned on $B(t_1) = a$ is normal with its mean equal to a and its variance equal to $t - t_1$.

Tied-down Brownian motion is Brownian motion conditioned on both the past and the future. In this circumstance, there are two times t_1 and t_2 such that $t_1 \leq t \leq t_2$. The values of $B(t_1)$ and $B(t_2)$ are known, that is, $B(t_1) = a$ and $B(t_2) = b$. In this case, the Brownian element may exhibit independent movement between these two points, but is fixed or tied-down at each of times t_1 and t_2 (Figure 6.6).

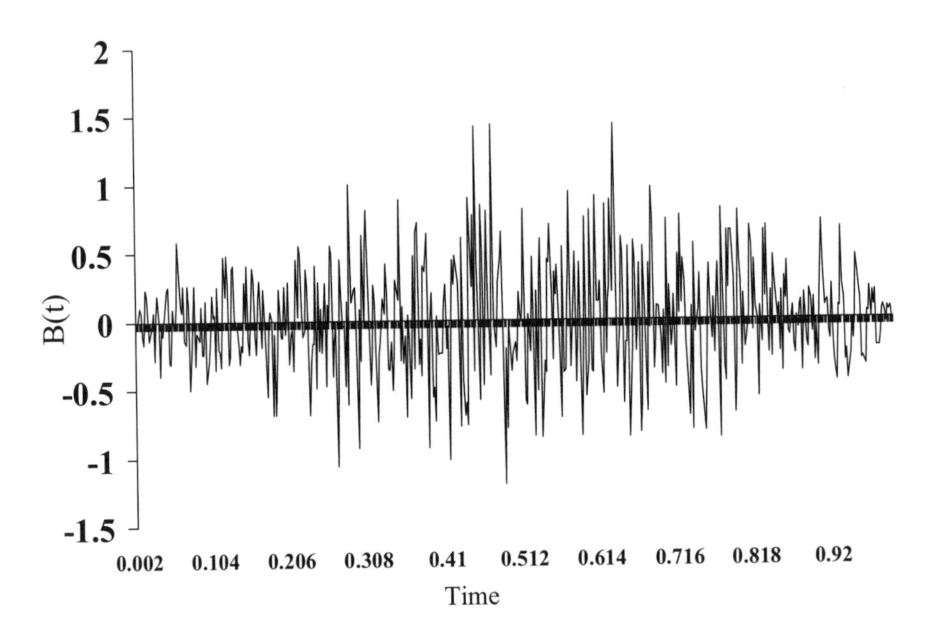

Figure 6.6. Brownian motion tied down to zero at t = 0 and 1.

The behavior of the Brownian element in Figure 6.6 exhibits the anticipated rapid random movement of Brownian motion. However, the extent of its excursions are limited. Because the value of $B(0) = B(1) = 0$, the variance of the movement is limited at these two points to zero. In fact, the variance of this tied-down Brownian motion increases from $t = 0$ to $t = 0.5$, and decreases thereafter (Figure 6.7).

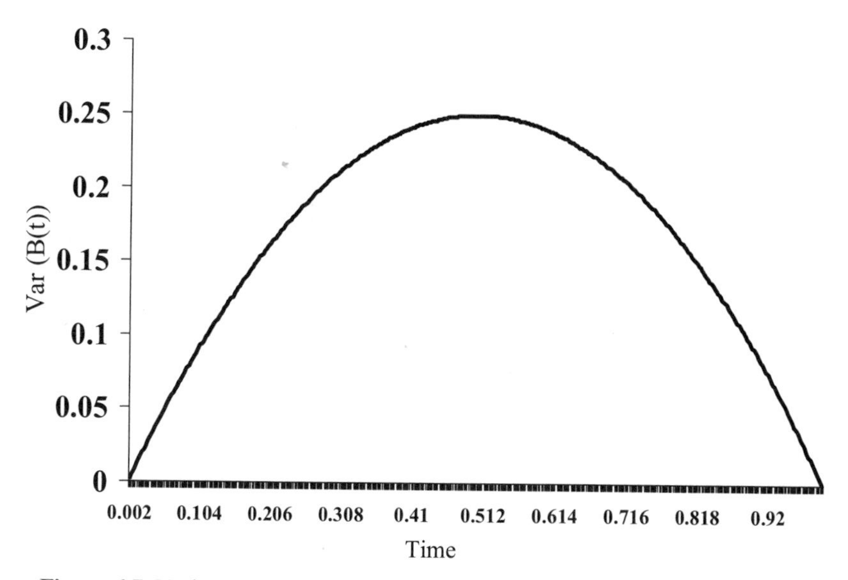

Figure 6.7. Variance of Brownian motion tied down to zero at $t = 0$ and $t = 1$ as a function of time.

In general, if $B(t)$ is Brownian motion tied-down at $t = t_1$ and $t = t_2$, such that $B(t_1) = a$, and $B(t_2) = b$, then it can be shown (Appendix C) that $B(t)$ follows a normal distribution with mean μ_B and variance v_B where

$$\mu_B = \frac{(t_2 - t)a + (t - t_1)b}{t_2 - t_1} = \frac{(t_2 - t)}{t_2 - t_1}a + \frac{(t - t_1)}{t_2 - t_1}b \tag{6.16}$$

and

$$v_B = \frac{(t_2 - t)(t - t_1)}{(t_2 - t_1)}. \tag{6.17}$$

Note that the mean of this process is a weighted sum of the two tied-down values of $B(t)$ were $t_1 \le t \le t_2$. When $t = t_1$, the summand in μ_B that includes b becomes zero, and $\mu_B = a$. As t is allowed to increase, the contribution of the term involving b increases until it becomes the major contributor to μ_B. Finally, at $t = t_2$, $\mu_B = b$. This behavior, in combination with the small variance of $B(t)$ at both $t = t_1$ and $t = t_2$, leads to the characteristic appearance of tied-down Brownian. Also note that when $a \neq b$, the Brownian bridge takes the drift parameter into account. (Figure 6.8).

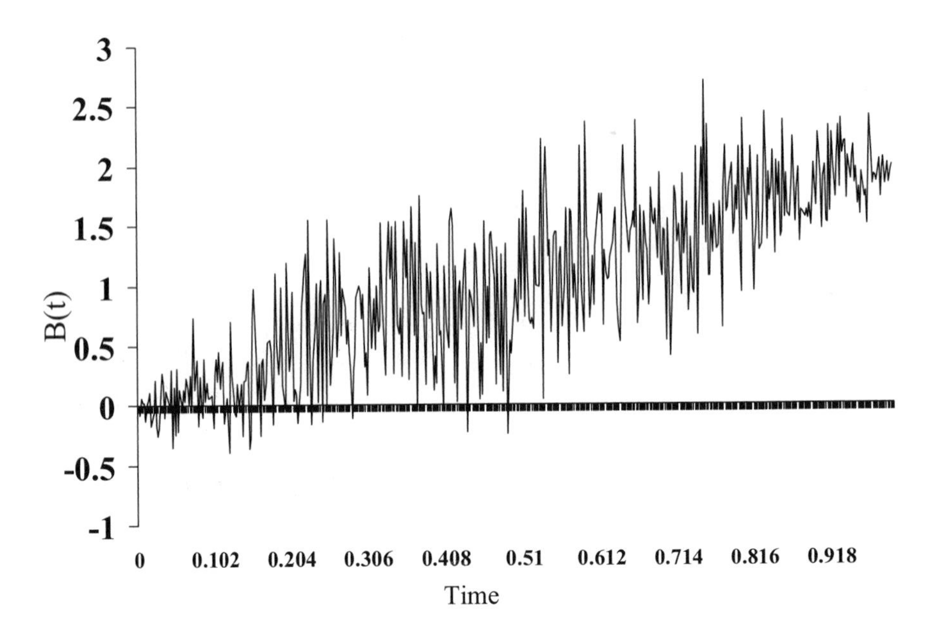

Figure 6.8. Tied-down Brownian motion such that $B(0) = 0$ and $B(1) = 2$.

The concept of tied-down Brownian motion would be useful to investigators who are interested in looking for deviations in the value of the test statistic from its trajectory. This trajectory can either be the desired one (that is, under the alternative hypothesis), or the anticipated one (that is, the past trajectory of the Brownian element). Deviations from the path can be taken as a warning sign that an unexpected path is emerging.

As an illustration, consider the original example of this section in which an investigator observes the value of the test statistic $TS(0.25) = 2.8$, and wishes to observe the expected trajectory of the test statistic. Specifically, she wishes to know if the movement is on a trajectory to attain the value of 4.45 when $I = 0.50$.

In this circumstance we note that the process is tied-down at the value 2.8 at $I = 0.25$ and the value 4.45 for $I = 0.50$. Information time will be allowed to increase in this range $0.25 \leq I \leq 0.50$. Writing the event in terms of Brownian motion, we convert the statement that $TS(0.25) = 2.8$ to an equivalent one involving a Brownian element, that is, $B(0.25) = \sqrt{0.25}\, 2.8 = 1.4$. Similarly, we can write $TS(0.50) = 4.45$ as $B(0.50) = 3.147$. The tied-down Brownian process is now calibrated. Its mean μ_B can be written as

$$\begin{aligned}\mu_B &= \frac{(I_2 - I)}{I_2 - I_1}a + \frac{(I - I_1)}{I_2 - I_1}b \\ &= \frac{(0.50 - I)}{0.50 - 0.25}1.4 + \frac{(I - 0.25)}{0.50 - 0.25}3.15 \\ &= 7I - 0.35\end{aligned} \quad (6.18)$$

The variance can be rewritten as

$$v_B = \frac{(I_2 - I)(I - I_1)}{(I_2 - I_1)} = \frac{(0.50 - I)(I - 0.25)}{(0.50 - 0.25)} = 4(0.50 - I)(I - 0.25). \quad (6.19)$$

This information allows us to build a 95% confidence interval around the Brownian process. Recall that if a variable x follows a normal distribution with mean μ and variance σ^2, then a 95% confidence interval for x, $CI(95)$, is $LB \le x \le UB$, where

$$\begin{aligned} LB &= u - 1.96\sigma \\ UB &= u + 1.96\sigma. \end{aligned} \quad (6.20)$$

Such a confidence interval can be identified for the Brownian element for every point I such that $0.25 \le I \le 0.50$. Using the equations from (6.20), incorporating the mean from equation (6.18), and the variance from equation (6.19), we can write

$$\begin{aligned} LB(I) &= 7I - 0.35 - 1.96\left[4(0.50 - I)(I - 0.25)\right] = 7.84I^2 - 1.12I + 0.63. \\ UB(I) &= 7I - 0.35 + 1.96\left[4(0.50 - I)(I - 0.25)\right] = -7.84I^2 + 12.88I - 1.33. \end{aligned} \quad (6.21)$$

For example, for the case where $I = 0.40$, the expected value of the Brownian process is 7(0.40) – 0.35 = 2.45, and the 95% confidence interval is (1.44, 2.57).

However, it would be most convenient to write these expressions not in terms of the Brownian element, but directly for the test statistic. Recall that the test statistic $TS(I)$ and $B(I)$ are related by $TS(I) = I^{-1/2}B(I)$, where I is a known constant. Thus, the mean μ_{TS} can be written as

$$\begin{aligned} \mu_{TS} &= I^{-1/2}(7I - 0.35) \\ v_{TS} &= 4I^{-1}(0.50 - I)(I - 0.25). \end{aligned} \quad (6.22)$$

This mean and confidence interval can be graphed as a function of I, for $0.25 \le I \le 0.50$ (Figure 6.9).

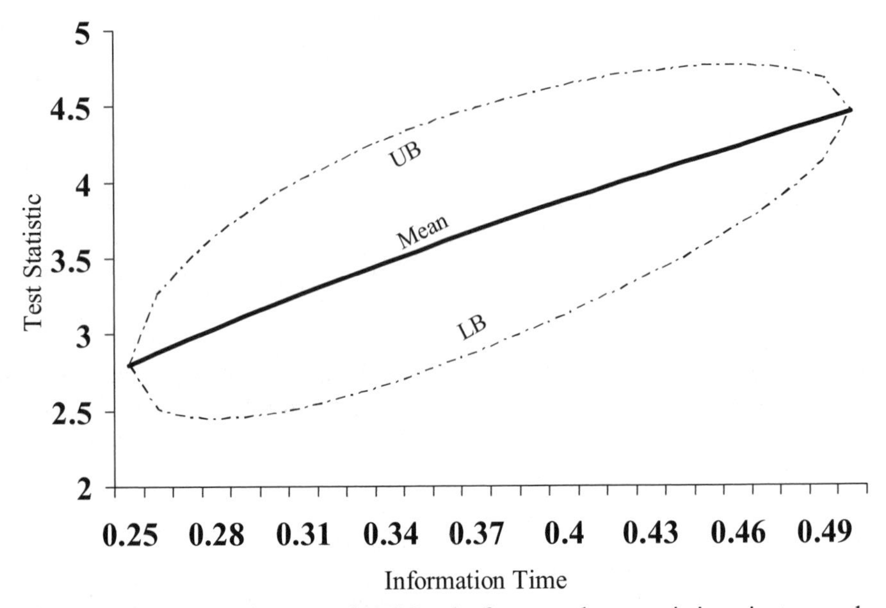

Figure 6.9. Expected mean value (Mean) of expected test statistic trajectory and its 95% lower bound (LB) and upper bound (UB) confidence interval.

We are now in a position to write the general result for tracking the trajectory of a test statistic across information time using tied-down Brownian motion. In general, let $TS(I_1)$ and $TS(I_2)$ be fixed for $0 \leq I_1 < I_2 \leq 1$. Then for any information time I such that $I_1 \leq I \leq I_2$, we can identify the mean $\mu_{TS(I)}$ and variance $v_{TS(I)}$ of TS(I) as

$$\begin{aligned} \mu_{TS(I)} &= I^{-\frac{1}{2}}\left[\frac{(I_2 - I)}{I_2 - I_1} TS(I_1) + \frac{(I - I_1)}{I_2 - I_1} TS(I_2)\right] \\ v_{TS(I)} &= \frac{(I_2 - I)(I - I_1)}{I(I_2 - I_1)}, \end{aligned} \tag{6.23}$$

and the lower bound (LB) and upper bound (UB) of the $1-\alpha$ confidence interval can also be computed as

$$\begin{aligned} LB &= I^{-\frac{1}{2}}\left[\frac{(I_2 - I)}{I_2 - I_1} TS(I_1) + \frac{(I - I_1)}{I_2 - I_1} TS(I_2)\right] - Z_{1-\alpha/2}\sqrt{\frac{(I_2 - I)(I - I_1)}{I(I_2 - I_1)}} \\ UB &= I^{-\frac{1}{2}}\left[\frac{(I_2 - I)}{I_2 - I_1} TS(I_1) + \frac{(I - I_1)}{I_2 - I_1} TS(I_2)\right] + Z_{1-\alpha/2}\sqrt{\frac{(I_2 - I)(I - I_1)}{I(I_2 - I_1)}}. \end{aligned} \tag{6.24}$$

6.9 Significance Levels and Early Termination*

Chapter Five and Chapter Six focused on the methodology of group sequential procedures to establish quantitative criteria to discontinue a study. The early termination of the study will produce for the investigators a test statistic and a *p*-value. The interpretation of these findings from a study that was terminated early can be problematic.

We have seen in Chapter Two that, even when permitted to proceed to its scheduled termination, a study's conclusions can be misleading when only the *p*-value of the result is conveyed. Difficulties in methodology can make the *p*-value u-interpretable, and small sample size, or the presence of important variability in the measure of effect can make reliance on the *p*-value misleading. For these reasons, the study methodology, sample size, confidence interval, and *p*-value must be interpreted jointly.

This perspective is easily translated to the early termination environment. However, there is a new question raised by the early conclusion of the study. Is the *p*-value of the study different from what it would have been if the study had been allowed to continue? An affirmative answer begs the question of whether the *p*-value produced by early termination should be discounted.

There are arguments on either side of this question. First, we must be clear on the circumstances. Any *p*-value that is produced by early termination when there was no prospective plan in place for interim monitoring is highly suspect, and most times should be discarded.* The role of sampling error in not just producing the data, but in creating the decision rule to stop the study when it was terminated is too distorting for the *p*-value to have any meaning in this circumstance.

However, when the *p*-value is produced from a clear, prospective, and competent interim monitoring plan resulting in early termination, a case can still be made for adjusting the *p*-value that is produced from the study. The issue in this case is not whether the study should have been terminated early. The concern is whether a *p*-value produced from a study that was concluded before its scheduled end is "equivalent" to a *p*-value that would have been produced if the study had been allowed to run its course. Some of the discussion has taken place in regulatory circles, where an argument has been made that the *p*-value for early termination of the study should be $p \leq \xi$, where ξ is the family wise error of the study for the one analysis. For example, if the α threshold at the conclusion of the study is 0.045, but the study was terminated when 75% information time for a *p*-value of 0.005, then the *p*-value of the study should be reported merely as less than 0.045.

However, this extreme argument itself leads to counterintuitive conclusions in some circumstances. Consider the hypothetical scenario of a clinical trial that following its prospectively declared, well-conceived interim monitoring plan, is ended at $I = 0.95$, with a *p*-value of 0.001 and conditional power of 95%. Should this *p*-value be adjusted up to $(0.05)(0.95) = 0.0475$ (a 475-fold increase) simply because it was concluded (5%) early?

---

* This is consistent with the requirement of the three principles delineated in Chapter Two for the prospective design of sample-based research in healthcare.

This reasoning suggests that if there is any adjustment to the p-value the extent of the adjustment might be best based on the unelapsed time of the trial, that is, a study that produces a test statistic leading to early termination at time I, $TS(I)$ should be adjusted for the un-elapsed information time in the study, $1 - I$. This idea might be pursued as follows. Let $p(I)$ be the p-value of the test statistic that leads to the termination of the study at information time I. Then define the p-value adjusted for early termination $p_T(I)$ as simply

$$p_T(I) = \left[p(I) - 1\right]I + 1. \tag{6.25}$$

When $I = 0$, the adjusted p-value is equal to 1. If the study is concluded as scheduled, then $I = 1$, and $p_T(I)$ is simply the original p-value. However, this function heavily discounts p-values that produce early termination. For example, a p-value of 0.007 that produces an early termination at $I = 0.50$, generates $p_T(I) = 0.504$!. This may seem like an extreme adjustment. However, an examination of the conditional power for early termination $Cp(H_0)$ produces

$$C_P\left(H_0\right) = \boldsymbol{P}\left[TS(1) \geq 1.96 \mid TS(0.50) = 2.7\right] = 1 - F_Z\left[\frac{1.96 - 2.7\sqrt{0.50}}{\sqrt{1 - 0.50}}\right]$$

$$= 1 - F_Z\left[0.719\right] = 0.236,$$

suggesting that there is indeed insufficient evidence to stop the study.

Problems

1. Let $TS(I)$ be a test statistic, such that $B(I) = \sqrt{I}\,TS(I)$ follows standard Brownian motion. Compute the following probabilities.

   a. $\boldsymbol{P}\left[TS(0.65) > 2.7 \mid TS(0.15) = 4\right]$.
   b. $\boldsymbol{P}\left[TS(0.35) \geq 2.7 \mid TS(0.15) = -2\right]$.
   c. $\boldsymbol{P}\left[TS(0.90) \geq 1.65 \mid TS(0.50) = 3.4\right]$.
   d. $\boldsymbol{P}\left[TS(0.90) \geq 2.1. \mid TS(0.80) = 3.1\right]$.

2. Let $TS(I)$ be a test statistic such that $B(I) = \sqrt{I}\,TS(I)$ follows Brownian motion with drift parameter = 2.034. Compute the following probabilities.

   a. $\boldsymbol{P}\left[TS(0.70) \geq 3.2 \mid TS(0.40) = 1.2\right]$.
   b. $\boldsymbol{P}\left[TS(0.50) \geq 1.5 \mid TS(0.05) = 3.9\right]$.
   c. $\boldsymbol{P}\left[TS(0.85) \geq 3.0 \mid TS(0.25) = 2.0\right]$.

d. $P[TS(0.40) \geq 2.1. | TS(0.30) = 2.1]$.

3. Suppose two clinical trials (clinical trial A and clinical trial B) are executed to answer the same scientific question by analyzing the same prospectively declared endpoint at the 0.05 level, unadjusted for multiple analyzes. Clinical trial A uses a conditional power interim monitoring procedure where $C_p(H_0) = 0.95$. Clinical trial B has no interim monitoring procedure. Each of the two studies ends with a test statistic of 1.97. Contrast the interpretation of the results of the two clinical trials.

4. Suppose two clinical trials (clinical trial C and clinical trial D) are executed to answer the same scientific question by analyzing the same prospectively declared endpoint at the 0.05 level, adjusted for multiple comparisons. Each uses a conditional power interim monitoring procedure where $C_p(H_0) = 0.95$. Clinical trial C has one interim look. Clinical trial D incorporates 10 interim looks. What is the difference in the type I error rate expended at each interim monitoring procedure for the two clinical trials? What are the other important interpretative differences between the methodologies used for these two clinical trials?

5. A clinical trial is designed to demonstrate that a clinical intervention can reduce the occurrence of the combined endpoint of fatal and nonfatal strokes by 30%. The cumulative event rate for fatal/nonfatal strokes in the control group is 12%. The two-sided alpha error rate for the single evaluation of the one primary endpoint at the conclusion of the study is 0.025 (adjusted for interim looks) to be carried out with 90% power. The trial size for this trial, (i.e., the total number of patients required for the study) is 3578, adjusted for interim evaluations. Fill in the following table (Table 6.3).

Table 6.3. Upper Boundary Values for Early Termination (Two Sided, Type I Error Rate = 0.0225 90% Conditional Power)

Information Time	Number of Events	Upper Boundary
0.25		
0.50		
0.75		
0.90		

6. Show that the drift parameter in tied-down Brownian motion such that $B(t_1) = B(t_2) = 0$, $t_1 \leq t \leq t_2$, is equal to zero.

7. Show that the drift parameter in tied-down Brownian motion such that $B(t_1) = a$, $B(t_2) = b$, $t_1 \leq t \leq t_2$, is equivalent to the drift parameter for tied-down

Brownian motion W(t) where $W(t) = B(t) - \frac{(t_1 a - t_2 b)}{(t_1 - t_2)}$, , where the drift parameter μ is $\mu = \frac{(b-a)t}{(t_2 - t_1)}$.

8. A researcher conducts a clinical trial that is governed by an interim monitoring procedure. At the scheduled monitoring time of I = 0.30, the test statistic is 3.89, producing a two-sided p-value of 0.0001. Compare and contrast $C_P(H_0)$ and $p_T(0.30)$.

References

1. Kyungmann K, and DeMets D. (1987) Design and analysis of group sequential tests based on the type I error spending rate function. *Biometrika* **74**:149-54.
2. Alling DW. (1966) Closed sequential testing for binomial probabilities. *Biometrika* **53**:73-84.
3. Halperin M, Lan KKG, Ware JH, Johnson NJ, DeMets DL.(1982) An aid to data monitoring in long-term clinical trials. *Controlled Clinical Trials* **3**:311-323.
4. Lan KK, and Wittes J. (1988) The B-value. A tool for monitoring data. *Biometrics* **44**:579-585.
5. Tsistis AA. (1982) Repeated significance testing for a general class of statistics used in censored survival analysis. *Journal of the American Statistical Association* **77**:855-861.
6. Grimmett GR, and Stirzaker DR. (1992) *Probability and Random Processes, 2nd ed.* Oxford. Clarandon Press.
7. Canner PL. (1977) Monitoring Treatment Differences in Long-Term Cancer Trials. *Biometrics* **33**:603-615.

7

Safety and Futility

7.1 Scope of This Chapter

At this point, we have discussed the implementation of modern statistical tools to monitor the interim results of clinical research for efficacy. There is an important ethical component to this imperative. The early termination of a clinical trial that has clearly established the effectiveness of an intervention can reduce the anticipated number of patients who have been exposed to the control group therapy, a therapy demonstrated to be inferior to the active intervention. The trial's early termination can also lead to the rapid dissemination of the study's positive result to the medical and regulatory communities. This practice spreads the knowledge, use, and benefits of the therapy to patients at large. These ethical concerns are reason enough to justify the monitoring of clinical research.

However, there are other products of clinical trial results that raise their own ethical concerns. Clinical research, at its heart, is an examination of the unknown. In medicine, the unknown, although frequently promising, can produce unforeseeable, but unmistakable harm. Thalidomide, and the two antiarrhythmic agents encainide and flecainide are examples of compounds whose blossoms of promise turned to sharp thorns of disappointment as agents intended to help have instead harmed. The long experience of healthcare research requires investigator-based vigilance for the possibility of the occurrence of avoidable injury. This requirement for the safe treatment of patients is both oath driven, and a clear expectation of the subjects enrolled in the study. These motivations combine to make safety the principal reason for monitoring clinical research.

Finally, the occurrence of a pattern of results that suggests that the study has produced neither a beneficial nor a harmful effect requires examination as well. The continued use of a therapy with perhaps minor, but nevertheless definable, adverse events in the absence of benefit can be difficult to justify ethically as well as economically. The idea of discontinuing a study because of the absence of an effect of the compound (described in the vernacular as stopping for *futility*) is recognized as yet another justification for terminating a study early and will receive special treatment in this chapter as well.

7.2 Complication in Safety Monitoring

The ethical concern for safety introduces a new complexity into the interpretation of interim monitoring results. Essentially, the primacy of the admonition, "First, do no harm," complicates the interpretation of exploratory analyzes. Recall that confirmatory analyzes, themselves the product of prospectively declared plans, are most easily generalizable. They produce reliable estimates of effect sizes, effect size variability, and type I error rates. In addition, prospective consideration of the type and number of statistical hypotheses to be tested leads to tight control of the familywise error rate. Exploratory analyzes, on the other hand, hold none of these features. These latter analyzes that are suggested by the data rather than by prospective planning produce inaccurate estimators and must be repeated (or confirmed) before they can be generalized.*

With this distinction in place, the evaluation of efficacy examinations in clinical research is clear and straightforward at both the conclusion of the study and during an interim examination. Consider a hypothetical clinical trial designed to determine the effect of an intervention on patients who have suffered a stroke. The primary analysis for this clinical trial is the effect of therapy on the total mortality rate. If the study's design and execution follows the three principles elaborated in Chapter Two, that is, the study has been (1) carefully considered with a precise, prospectively declared protocol, (2) executed in accordance with that protocol,† and (3) produces a type I error rate below the a priori declared threshold for a clinically significant level of efficacy, its results on the primary endpoint would be heralded as positive. However, assume in that same study, that the therapy was observed to produce a reduction in the myocardial infarction rate. Because the therapy's effect on the myocardial infarction rate was not prospectively declared, and its results would be seen as exploratory‡, requiring confirmation before they can be generalized.

However, in the reverse setting, where both the findings for mortality and the myocardial infarction rate indicate harm and not benefit, the ability to interpret the situation is complicated. Again, the prospective nature of the research design

---

* The reasons for the different interpretations of exploratory versus confirmatory analyzes are discussed in Chapter Two.

† This is defined as concordant execution.

‡ Because both the effect of therapy on the total mortality rate and the effect of therapy on the myocardial infarction rate appear to be equally reliable to the observer, both being produced from the same database, concluding that only the myocardial infarction finding be discounted may appear capricious. However, because the trial was not designed to examine the myocardial infarction issue, the ability of the study to ensure that (1) it had an appropriate sample size for the evaluation of the effect of therapy on heart attacks, and (2) the investigators collected all of the cases of heart attacks (including silent myocardial infarctions) even though they never said that they would must be questioned. Secondly, the "positive" finding of therapy on the myocardial infarction rate was based on an unplanned interrogation of the database. This inquiry rose to prominence solely because of its unanticipated and surprising finding. Allowing the data to suggest answers in this fashion introduces a new sampling error component into the analysis that undermines the traditional estimates of effect size, precision, p-values and power.

that is focused on mortality permits the investigators to conclude that the study produces a confirmatory evidence of harm. However, the identification of an early harmful effect of the intervention on the cumulative myocardial infarction rate is problematic. Clearly, the finding of an increased risk of myocardial infarction associated with the therapy is crippled by the exploratory nature of the analysis. However, the occurrence of this finding for harm in all likelihood would not, and could not, be reproduced. The ethics of protecting subjects from known and suspected dangers precludes the initiation of a new confirmatory research effort in order to demonstrate that the intervention excites the production of myocardial infarctions.*

Thus, although confirmatory findings of harm carry the same weight as those of efficacy, exploratory findings of harm commonly carry more weight than exploratory findings of efficacy. This places a greater burden on investigators, who must balance the ethical requirement to adequately warn about adverse events with the scientific need to avoid misdirection about the risks of an intervention. This is a challenge to healthcare research in general and interim monitoring in particular. Because, confirmatory analyzes for harm will be the most persuasive, investigators make their strongest possible case when they execute as many confirmatory analyzes as possible.

Although we will suggest two strategies to address this issue of multiple safety measures, the best strategy begins with good a priori knowledge. During the design phase, investigators can strengthen the interim analyzes that they have in mind for the safety monitoring by carrying out an in-depth review of the evidence of adverse effects produced by the intervention that they plan to study. This will allow them to put prospective monitoring rules in place that provide rigor in detecting the early occurrence of several different adverse events.

7.3 Monitoring Safety for Primary Endpoints

We begin with the simplest case for monitoring safety in a clinical trial: the creation of a guideline for early termination triggered by the occurrence of harm appearing in the primary analysis of a study. Recall from Chapter Two that primary analyzes have two defining characteristics: (1) the well-designed evaluation is described a priori, and (2) a type I error rate is prospectively conferred in such a way as to minimize the overall or familywise type I error rate. In this circumstance, our previous discussion of group sequential procedures readily applies to this setting, and investigators carrying out a two-tailed hypothesis test can easily construct a monitoring procedure for harm.

7.3.1 Symmetric Boundaries

As an example, consider a prospective, randomized, double-blind clinical trial designed to determine the effect of an anticoagulant therapy on the total mortality rate of patients with coronary artery disease. Patients are followed for two years in the study. The evaluation at the conclusion of the study will be a two-tailed evaluation,

* Other information may become available that would allow the evaluation of the possibility of harm produced by the intervention. Additional concurrent studies, case control, and historical cohort studies may provide some illumination of this issue.

and the investigators will monitor for both a beneficial or harmful effect of mortality during the interim examination times at $I = 0.10, 0.30, 0.50, 0.75$. We will develop the boundary values for benefit and for harm using the conditional power approach, with conditional power $\gamma = 0.90$.

Recall from Appendix C that the boundary values for efficacy during the interim examination period can be written as

$$b_e(I) = \frac{Z_{1-\alpha/2} - Z_{1-\gamma}\sqrt{1-I}}{\sqrt{I}}, \quad (7.1)$$

where $b_e(I)$ denotes the boundary value for efficacy at information time I. Use of the percentile value $1-\alpha/2$ reminds us of the two-sided nature of test. The event that illuminates the possibility of harm is the probability that the test statistic is "too negative", that is, it falls into the lower tail of the probability distribution, given that the test statistic at information time I is equal to b. This may be written as $\boldsymbol{P}\left[TS(1) < Z_{\alpha/2} \mid TS(I) = b\right]$. Appendix C also revealed that the boundary for harm is

$$b_h(I) = \frac{Z_{\alpha/2} - Z_{\gamma}\sqrt{1-I}}{\sqrt{I}}, \quad (7.2)$$

where $b_h(I)$ denotes the boundary value for harm at information time I. If the conditional power for benefit is equal to that of harm, then $b_h(I) = -b_e(I)$ and the boundary region is symmetric. This is illustrated in the current example in which the conditional power is 90% and the familywise error rate at the conclusion of the study is $(0.90)(0.05) = 0.045$ (Table 7.1).*

If we take the point of view that, *ceteris parabus*, the magnitude of the test statistic provides an indication of the magnitude of efficacy, then the use of symmetric boundaries signals that the investigators require equal strength of evidence for identifying benefit as well as harm.

---

* This computation is simply the adjustment in the final type I error of the trial given that interim monitoring will take place. Its motivation was discussed in Chapter Six.

Table 7.1. Symmetric Boundary Values for Early Termination (Two-Sided, Type I Error Rate = 0.045, 90% Conditional Power)

Information Time	Upper Boundary	Lower Boundary
0.10	10.18	-10.18
0.30	5.62	-5.62
0.50	4.12	-4.12
0.75	3.05	-3.05
1.00	2.00	-2.00

The depiction of these boundaries graphically reveals the type of curves that have come to symbolize the use of group sequential procedures in the statistical monitoring of clinical research (Figure 7.1).

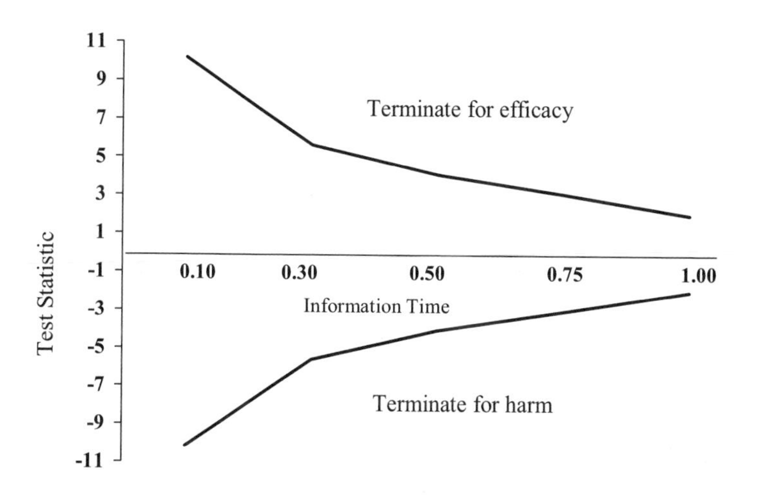

Figure 7.1. Conditional power monitoring guidelines with equal consideration given to the early appearance of efficacy or harm.

In this circumstance, both extreme positive values and extreme negative values of the test statistic are cause for concern and can precipitate discussions for early trial termination.

7.3.2 Asymmetric Boundaries

The previous treatment demonstrates the ease of developing boundary conditions for harm as well as for benefit that can be used in the interim monitoring of clinical research. However, there is no mathematical requirement for the use of symmetric boundaries, and boundaries that provide less conditional power for harm can be readily generated. Consider the example in the previous section, now developed requiring 95% conditional power for benefit, and 80% conditional power for harm (Table 7.2).

Table 7.2. Asymmetric Boundary Values for Early Termin (Type I Error Rate for Benefit = 0.0237, 95% Conditional Benefit; Type I Error Rate for Harm 0.020, 80% Conditior for Harm).

Information Ti	Upper Bou	Lower Boundary
0.10	11.20	-9.02
0.30	6.13	-5.04
0.50	4.45	-3.75
0.75	3.24	-2.86
1.00	1.98	-2.05

There are several interesting observations available from Table 7.2. First, we must acknowledge that the boundary values are less extreme for harm than for efficacy. Although this is an anticipated outcome from the asymmetrical assumption that is the foundation of Table 7.2, it is also true that the important asymmetry in conditional power, 95% for efficacy, now 80% for harm, did not produce overly moderate boundary values for assessing the presence of harm. Even with a $(0.95-0.80)/0.95 = 15.7\%$ reduction in conditional power, the boundary values for stopping for harm remain substantial (Figure 7.2). The reduction in conditional power was not so great as to overshadow the unreliability of early estimates of mortality (and therefore, the test statistic that is based on mortality). Thus, large (negative) values of the test statistic are still required early in the study in order to instigate conversation about terminating the study early for the presence of an adverse finding for the primary endpoint.

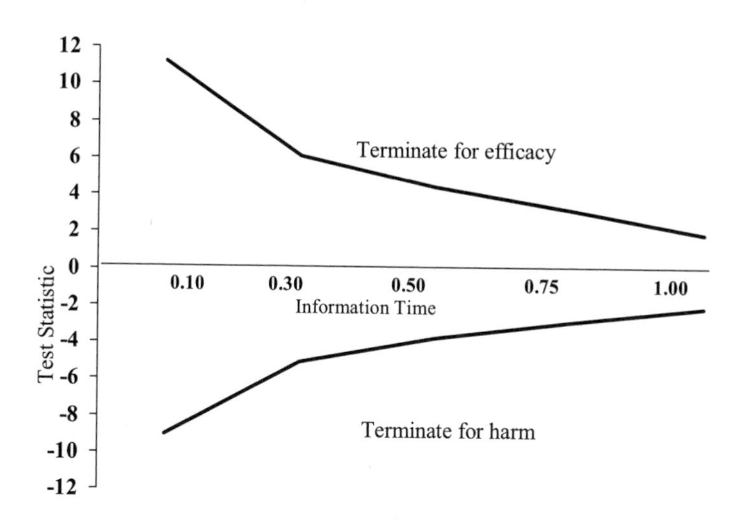

Figure 7.2. Asymmetric conditional power monitoring guidelines. 95% conditional power for benefit and 80% conditional power for harm for the primary endpoint.

7.3.3 Multiple Prospective Safety Endpoints*

The computation for the boundary values for the value of a test statistic for the interim monitoring of the efficacy and safety of a single, prospectively declared primary endpoint is relatively straightforward. However, this is a fairly rare circumstance in clinical research. Commonly, although the investigator must acknowledge that one safety consideration is the demonstration of harm produced by the intervention, a second and more complex problem is that other adverse events may be produced by the intervention being evaluated. We will evaluate this circumstance by first assessing how one can monitor multiple prospectively declared adverse events.

When an intervention is considered for formal study in a clinical trial, the therapy has already established at least the suggestion of a side-effect profile. Knowledge of this side effect profile is necessary in order to provide the best treatment for these patients in the study. For example, if patients in a clinical study are likely to experience blurred vision as a consequence of the intervention, then the trial mechanism can incorporate frequent examinations for the occurrence of blurred vision. In addition, adjunct therapy can be considered to alleviate the impact of this troublesome adverse effect.

Similarly, foreknowledge of the occurrence of a serious adverse event profile can direct the monitoring plan of a clinical study. Understanding the risk profile of an intervention permits the researchers to tailor the monitoring plan of the study to the magnitude of the intervention's risk that patients must bear. Because the investigators can only justify the use of the intervention when its benefits outweigh its risks, the monitoring procedure can be sculpted to be a manifestation of this risks, the monitoring procedure can be sculpted to be a manifestation of this risk–benefit balance. This realization can be constructively framed within the confines of type I error consideration and conditional power.

Consider the following example. Clinical trial investigators are interested in investigating the effect of an intervention that is believed to reduce the stroke mortality rate. Subjects are recruited into the study while experiencing a stroke, and then, in addition to being provided the state-of-the-art stroke treatment, are randomized to receive either the intervention or placebo therapy. In this study, the investigators measure the severity of the stroke by assessing the patients' abilities to effectively control their movements and express themselves. This assessment is captured in the National Institutes of Health Stroke Scale (NIHSS). A patient's capabilities are graded, and points are given for the abnormalities observed in accordance with pre-stated criteria. A score of 0 reflects no difficulty at all, whereas patients with profound post-stroke disability can have stroke scores above forty.

The investigators wish to demonstrate that the therapy in which they are interested limits the stroke size in patients assigned to the active group. Thus, the researchers want to track the trajectory of the NIHSS stroke score from the baseline measurement (taken when the patient enters the hospital) to 28 days after the stroke has occurred. It is their hope that the change in NIHSS score will lead to greater improvement in the patients assigned to the intervention than in patients assigned to placebo therapy.

However, the investigators also recognize that the intervention is associated with serious adverse events. One of these is internal bleeding. The second is acute liver injury, as measured by dramatic rises in liver enzyme levels. Although the need for the most precise estimates of these adverse events for these patients would require greater, rather than fewer patient numbers, the investigators recognize the need to capably and ethically monitor the study, perhaps terminating it early if the therapy appears unsafe. Based on the philosophy that, if the therapy is to be worth its risks, then the potential for benefit must be greater than the potential for internal bleeding or liver damage, the investigators allocate type I error during the design of the trial in this risk-averse research climate (Table 7.3).

Table 7.3. Allocation of Familywise Error for Benefit and Harm

Endpoint	Test-Specific Alpha	Familywise Error	Conditional Power	Adjusted Test Specific
Benefit				
Stroke scale	0.025		0.95	0.024
		0.025		
Harm				
Stroke scale	0.025		0.50	0.013
Internal bleeding	0.150		0.50	0.075
Liver injury	0.100		0.50	0.050
		0.275		

The investigators demonstrate through the design of this study (as portrayed in Table 7.3) the importance of providing adequate protection for patients against the harm of serious adverse effects. The familywise error is the type I error allocated to the interpretation of the statistical hypothesis test at the conclusion of the study. Although it is appropriately low for benefit (at the 0.025), it is high (more than ten times higher) for the identification of a harmful effect.

Examining the finding for serious adverse events in greater detail, observe that for each of the types of the three serious adverse events (stroke scale, internal bleeding, and liver injury) the test-specific alpha error rate is high, revealing the sensitivity of the investigators to the occurrence of these serious adverse events. It is these large selections for the test-specific alpha error rates for serious adverse events that produce the high familywise error rate for harm of 0.275. Essentially, the investigators are willing to accept a large type I error rate in order to ensure that they do not miss the occurrence of a harmful effect that can occur in the population at large.

Proceeding across Table 7.3, we observe that the investigators take an additional step to ensure that the monitoring rule is appropriately attuned to the possibility of the occurrence of harm. We would not be surprised to observe the high conditional power for efficacy. However, the prospectively declared, conditional power choices of 50% are quite low, again reflecting the investigators desire to go out of their way in identifying a potentially important relationship between the occurrence of these serious adverse events and the therapy.

The boundary values produced from these computations are instructive (Table 7.4). Specific boundary values are computed for each of the one efficacy measure and the three harm measures. The boundaries are endpoint-specific, because each endpoint has its own test-specific alpha error and its own conditional power.

Table 7.4. Boundary Values for Multiple Endpoints

Endpoint	Test-Specific Alpha	Conditional Power	Boundary Values Information Time				
			0.10	0.30	0.50	0.75	1.00
Benefit							
Stroke scale	0.025	0.95	11.20	6.13	4.45	3.24	1.98
Harm							
Stroke scale	0.025	0.50	-7.09	-4.09	-3.17	-2.59	-2.24
Internal bleeding	0.150	0.50	-4.55	-2.63	-2.04	-1.66	-1.44
Liver injury	0.100	0.50	-5.20	-3.00	-2.33	-1.90	-1.64

Each of the endpoints for harm is allocated its own familywise error rate and conditional power

A graphic depiction is also revealing (Figure 7.3).

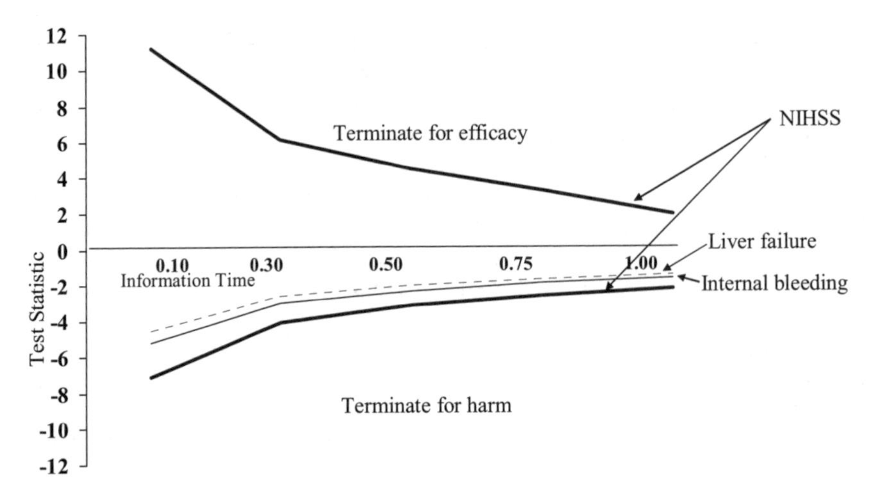

Figure 7.3. Asymmetric conditional power monitoring guidelines for multiple endpoints for harm (NIHSS, internal bleeding, liver failure). Alpha errors and conditional power are each test-specific in accordance with Table 7.4. The familywise error for benefit is 0.025 and the familywise error for harm is 0.275.

In their attempt to balance the role of sampling error and the occurrence of these adverse events, the investigators accept the risk of making a type I error to identify a risk of serious adverse events in the population.

It is also important to note that this type of computation is governed by parameters that the investigators select. Specifically, the researchers choose the test-specific type I error rates and the conditional power for the endpoint monitoring.

7.3.4 Combined Safety Endpoints*

Typically in designing a clinical research effort, investigators will choose an endpoint that is a composite of several different but related endpoints. This type of endpoint is a *combined endpoint* or a *composite endpoint.* For example, a clinical research effort may wish to focus on the occurrence of fatal and nonfatal stroke. In this circumstance, every patient who has had a fatal stroke, as well as patients who survive but have a nonfatal stroke are counted as having endpoints.

The advantages of this approach are well described [1]. A carefully constructed combined endpoint can helpfully broaden the definition of a clinical endpoint when the disease being studied has different clinical consequences. This expansion commonly increases the incidence of the endpoint, reducing the sample size of the trial. Alternatively, the use of a combined endpoint can increase the sensitivity of the research effort to detect moderate levels of therapy effectiveness.

Typically, these combined endpoints are constructed with the view that the therapy will be beneficial, producing a positive effect on the occurrence of the combined endpoint. However, there is no methodological barrier to constructing a combined safety endpoint. The advantage of a combined safety outcome is that sensitivity to the occurrence of dangerous trends in the safety data can be enhanced with tighter control of the familywise error rate.

As an example, consider the effect of an intervention that is designed to reduce coronary artery vasoconstriction. The intervention must be delivered intravenously. The primary endpoint for this study is the occurrence of fatal or nonfatal myocardial infarction. The researchers are concerned about the adverse effects of this intervention on the lungs. The investigators suspect that the therapy can produce a collection of nonserious but troublesome respiratory adverse effects. However, they are also concerned about the occurrence of more serious adverse events that affect the lungs.

In their attempt to monitor for the occurrence of pulmonary adverse events, the investigators choose to create two combined endpoints. The first, termed P_m, is designed to capture the first occurrence of either of the following mild adverse events: (1) complaint of a new, dry cough, (2) mild but non-limiting wheezing. The second combined endpoint, termed P_s combines the occurrence of serious adverse events, for example, dyspnea, hemoptysis, acute bronchitis, and hospitalization for pulmonary illness (Table 7.5).

The investigators are interested in monitoring the occurrence of these events for the possibility of harm.

Table 7.5. Combined Endpoints for Safety Monitoring

Combined endpoint	Components
Combined endpoint P_m	Dry Cough Nonlimiting wheezing
Combined endpoint P_s	Dyspnea Hemoptysis Acute bronchitis Hospitalization for pulmonary event

For each of these endpoints, the investigators are interested in computing boundary values for the test statistic that reflects the effect of therapy for each of them (Table 7.6).

Table 7.6. Boundary Values for Multiple Safety Endpoints

Endpoint	Test-Specific Alpha	Conditional Power	Boundary Values Information Time 0.05	0.25	0.50	0.70	1.00
Benefit							
Fatal/nonfatal MI	0.025	0.90	10.18	5.62	4.12	3.05	2.00
Harm							
Fatal/nonfatal MI	0.025	0.50	-7.09	-4.09	-3.17	-2.59	-2.24
P_m	0.010	0.99	-14.35	-7.81	-5.62	-4.03	-2.33
P_s	0.250	0.50	-3.64	-2.10	-1.63	-1.33	-1.15

MI = myocardial infarction

Table 7.6 provides the test-specific alpha, conditional power, and resulting boundary values for measuring the boundary values based on information time for monitoring the clinical events. There is one primary endpoint that maintains the efficacy focus. The boundary values for it are as anticipated, given the test-specific alpha and the high conditional power.

In addition, there are three monitored endpoints for harm. The first is of course the primary endpoint, fatal/nonfatal myocardial infarction. The second is the combined endpoint for safety, assessed with a test-specific alpha level of 0.01, and 99% conditional power. The low test-specific alpha and high conditional power for P_m suggest that only hyperextreme findings for the occurrence of this endpoint, measuring mild pulmonary function disease, would lead to the trial's early termination. The limiting, sometimes tragic, consequences of the adverse events that comprise the occurrence of P_s suggest that this safety endpoint be very carefully and sensitively monitored. This sensitivity is transmitted through the high test-specific alpha (0.025) and low (50%) conditional power (Figure 7.4).

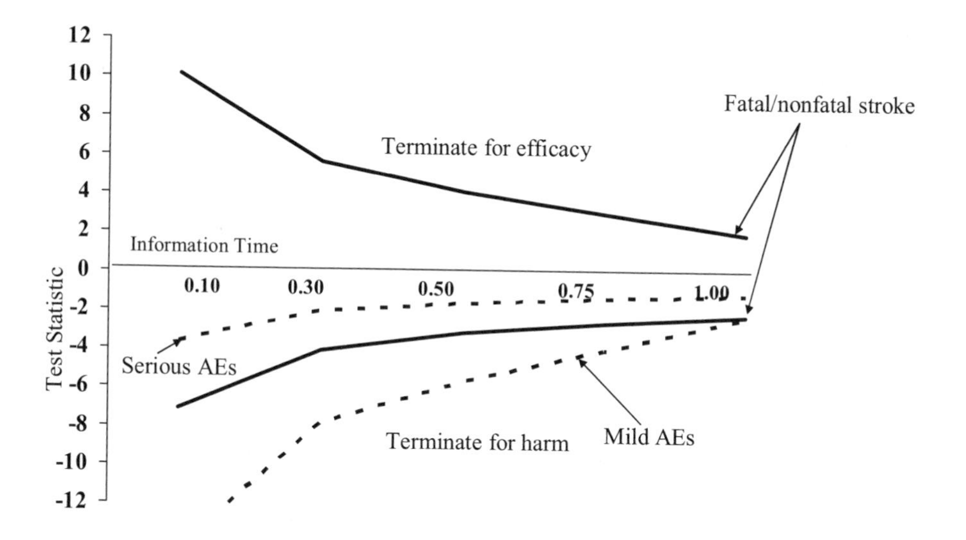

Figure 7.4. Asymmetric conditional power monitoring guidelines for multiple combined endpoints. Alpha errors and conditional power are each test-specific in accordance with Table 7.6.

7.4 Monitoring Safety in Small Samples*

Clinical research involving small samples commonly presents conundrums for monitoring adverse events. Contemporary IRBs are as focused to the safe conduct of these studies that commonly enroll relatively small numbers of patients as they are in the much larger clinical efforts. However, small clinical studies can make the group sequential procedure unwieldy. Data may not arrive in groups, but instead becomes available on a patient-by-patient basis.

A common scenario in these situations is the following. An investigator is focused on the occurrence of an adverse event rate in a series of patients in whom he is studying an IRB-approved intervention. He knows the background rate of the adverse event. His major safety concern is whether the occurrence of the adverse event in his study is exceeding the background rate. Specifically, he wishes to learn how many additional patients he must observe in order to have some assurance that the event rate is greater than the background rate. If these additional patients are a small number, he has enough information to terminate the study.

For example, an investigator is studying a new intervention for stroke, but for which bleeding is an important adverse event rate. The background rate for an intracranial hemorrhage is 6%. She has observed 100 patients in her study so far, and, in this cohort, 3 patients have experienced this event. The IRB has approved the recruitment of an additional 50 patients for the study. How many patients with a severe bleeding event must occur in this subsequent cohort of 50 patients in order to be confident that the bleeding rate in the entire study cohort is greater than the background rate of 6%?

The evaluation is straightforward. Let the underlying background rate of the internal bleeding be denoted by p_B. In her cohort, the investigator has observed n_0 patients thus far, of which k_0 patients have experienced an adverse event. The study will recruit another n_1 patients. We need to find the number k_1 of patients who, if they have an internal bleeding event in this second cohort of n_1 patients will boost the study adverse event to the level that the investigator can feel confident that the study's rate is greater than the background rate. Then at the conclusion of the next segment of the study, the cumulative event rate observed in the study is p where

$$p = \frac{k_0 + k_1}{n_0 + n_1}.$$

The concern is whether p is greater than the background rate (denoted by p_B) of the adverse event. We can use the confidence interval to help determine how much greater the study rate should be from the background rate. Specifically, the investigator would be alarmed if $p - p_B$ fell outside of its $1 - \alpha$ confidence interval. If we denote $Z_{1-\alpha}$ as the $1 - \alpha$ percentile value of the normal distribution, then p is extreme if

$$p - p_B \geq Z_{1-\alpha}\sqrt{\frac{p_B(1-p_B)}{n_0 + n_1}}$$

or

$$p \geq p_B + Z_{1-\alpha}\sqrt{\frac{p_B(1-p_B)}{n_0 + n_1}}.$$

Substituting $p = \frac{k_0 + k_1}{n_0 + n_1}$. into this last expression, we find

$$\frac{k_0 + k_1}{n_0 + n_1} \geq p_B + Z_{1-\alpha}\sqrt{\frac{p_B(1-p_B)}{n_0 + n_1}},$$

and we can solve for k_1

$$k_1 \geq \left[p_B + Z_{1-\alpha}\sqrt{\frac{p_B(1-p_B)}{n_0 + n_1}} \right](n_0 + n_1) - k_0.$$

This provides the smallest number of adverse events that we can tolerate in the next n_1 patients that will provide evidence that the rate of adverse events in the study is greater than the background rate. For example, suppose that the investigator anticipates the baseline rate for internal bleeding is 6%, and 3 patients have been observed in 100 patients, with the study poised to recruit an additional 50 patients. Then, in this circumstance $p_B = 0.06$, $k_0 = 3$, $n_0 = 100$, $n_1 = 50$, and $Z_{0,975} = 1.96$. The minimum value of k_1 is

$$k_1 \geq \left[0.06 + 1.96\sqrt{\frac{(0.06)(0.94)}{100+50}} \right](100+50) - 3 = 11.3.$$

In this example, if 12 patients out of the next 50 experience an ICH, then the total event rate is $\frac{(3+12)}{(100+50)} = 15/150 = 0.10$. The number of adverse events can be counted in the next 50 patients, and if it exceeds 12, termination activity should be considered.

A small sample procedure is available if the size of the trial is deemed too small to apply the normal distribution. For example, an investigator is carrying out an uncontrolled study evaluating the safety of an intervention. Specifically, she needs to assure herself that the rate of adverse events observed in this trial is not greater than that experienced in the population not subjected to the intervention.

Given that there are anticipated to be N patients recruited and followed in the study, in which A of them will have adverse events. The accepted rate of adverse events is anticipated to be p_0. Thus $A = Np_0$ is the expected number of patients anticipated to have adverse events at the trial's conclusion. However, at the current time in the study there are $n(k)$ patients recruited, k of which have experienced an adverse event. The investigator needs to know the probability that the total number of adverse events at the study's conclusion exceeds Np_0, given the current results, or

$$\boldsymbol{P}\left[A \geq Np_0 \mid n(k) = k\right].$$

If we assume that the number of adverse events A follows a binomial distribution, we may compute this probability directly. Because $n(k)$ patients have already been enrolled, producing k adverse events, then in the remaining $N - n(k)$ patients yet to be observed, there must be greater than $Np_0 - k$ adverse events to exceed the threshold. However, the number of remaining adverse events in $N - n(k)$ also follows a binomial distribution, thus, we may write

$$\boldsymbol{P}\left[A \geq Np_0 \mid n(k) = k\right] = \boldsymbol{P}\left[A - n(k) \geq Np_0 - k\right] = \sum_{j=Np_0-k}^{A-n(k)} \binom{A-n(k)}{j} p_0^j \left(1-p_0\right)^{A-n(k)-j}.$$

For example if there are 3 patients with adverse events in 7 patients, and the study will recruit 50 patients with an adverse event rate of 10%, then the probability that this event rate will be exceeded is

$$\sum_{j=Np_0-k}^{A-n(k)} \binom{A-n(k)}{j} p_0^j \left(1-p_0\right)^{A-n(k)-j} = \sum_{j=2}^{43} \binom{43}{j} (0.10)^j (0.90)^{A-n(k)-j} = 0.818.$$

Thus there is an 82 percent chance that the number of adverse events at the conclusion of the trial will exceed 5.

7.5 Exploratory Monitoring for Safety

The previous two sections described processes by which prospectively declared rules for monitoring adverse events can be generated. These rules can be sensitively calibrated to the concerns of the investigator who wishes to understand the risks and benefits of the therapy under consideration. However, that entire discussion's foundation was the prospective identification of risks and benefits. With foreknowledge of the adverse event profile of the intervention, and some appreciation of the frequency at which these adverse events appeared, the researcher is able to place prospectively declared rules in place.

However, many times adverse events occur that were completely unsuspected during the course of a clinical research effort which catch the researcher by surprise. The occurrence of de novo breast cancer in women exposed to lipid lowering agents during the course of a cardiovascular clinical experiment, the occurrence of dangerous new arrhythmias generated by an intervention that suppresses other rhythm disturbances, and the production of liver failure in medications designed to reduce insulin sensitivity are but a few examples of unanticipated adverse effects that can be produced by a new intervention.

This illustration suggests that although monitoring procedures can provide clear warning that a treatment signal has been detected, the presence of this signal does not provide prima facie evidence for early termination of the research effort. That decision is more complex, requiring thoughtful consideration of both the efficacy and the safety of the compound. Ultimately, the decision to discontinue a study is not a mathematical one, but a clinical and ethical decision.

The use of viruses as vectors for the transformation of the patient's genome with a new DNA strand multiplies the concern for adverse events. The new genome with its altered DNA sequence can conceivably produce many unanticipated effects on organ tissue and germ cells. The concern for safety in this new circumstance predominates, but one doesn't know where the adverse events will appear, or whether they will appear at all.

Statistical monitoring rules do not provide much guidance in this scenario. The absence of prospective identification, in concert with the large probability of a type I error generated by multiple interrogations of the adverse event data confound any attempt to apply statistical rigor to the evaluations executed on behalf of the patients. This is the conundrum of safety-based monitoring. The predomination of ethical considerations requires the researcher to sometimes set aside statistical precepts in order to be assured that patients are not being harmed in the research effort or in the population. This is unacceptable when one is building an argument for efficacy, but is required in the examination of safety. The asymmetry of this circumstance finds its genesis more in an oath than in science.

The following are other useful tools in assessing the validity of an unanticipated adverse event finding.

7.5.1 Monitor Widely

Researchers must lead diligent adverse event report evaluations. Just as physicians cannot diagnose diseases of which they have never heard, researchers cannot ethically weigh the implications of serious adverse events that they are not measuring. Thus, the researchers must cast a wide net in order to capture the occurrence of all adverse events.

The National Institute of Neurologic Disorders and Strokes (NINDS) requires that each DMC-monitored trial have an assigned Medical Safety Monitor (MSM) who bears responsibility for the review of individual serious adverse events (SAEs). This monitor must not be an investigator, and report these adverse events to DMCs on a regular basis. In clinical trials, the MSM should be a physician who is not involved in the study and who has no conflict of interest. They can suggest protocol modifications to prevent the occurrence of particular adverse events (e.g., requiring more frequent measurement of laboratory values).

The MSM will prepare regular reports concerning SAEs (not grouped by therapy assignment) for the principal study investigator, the DMC, or, where appropriate, other agencies (e.g., the FDA and collaborating biopharmaceutical companies or device manufacturers).

Should unexpected or an unanticipated high rate of SAEs occur, the adverse event monitor must immediately notify the principal study investigator and the DMC.

7.5.2 Assessing Unanticipated Adverse Events

We have focused on the statistical monitoring of clinical research. However, we saw in Chapter One that these quantitative tools have been in use for approximately sixty years, (from the earliest work of Wald [2]). This is a relatively short period of time in the history of clinical investigation. However, the assessment of whether the occurrence of the event is related to a treatment has been a cerebral process that has evolved for hundreds of years. Useful skills have been amassed by scientists in the implementation of nonstatistical evaluations of the relationship between an event (be it either adverse or salubrious) and an exposure. In the setting of exploratory endpoint analyzes, where statistical assessments are not quite so reliable, these other nonstatistical assessments rise to new prominence.

The occurrence of an exposure and an adverse event occurring in the same patient (or group of patients) begs the question of whether the relationship between the two is associative or causal. An associative relationship is merely that the exposure and the disease occurred coincidentally in the same patients, for example, the occurrence of breast cancer in women taking HMG-CoA reductase inhibitor (i.e., statin) therapy for elevated lipid levels. It is a passive relationship. On the other hand, a causal relationship is active. An exposure causes a event if the exposure excites the production of the event. It is an active relationship, a relationship with directionality. An example is the causal relationship between liver hepatotoxicity and thiazolidienedione therapy for diabetes mellitus.

When statistical tools are less useful, other tools of observation become more valuable. In 1965, Hill [3] described the nine criteria for causality arguments in healthcare. These nine rules or tenets are remarkably and refreshingly devoid of complex mathematical arguments, relying instead on natural honest intuition and common sense for the inquiry into the true nature of a risk factor–disease relationship. The questions Dr. Hill suggested should be posed by the investigators in their assessment of the true nature of the exposure–disease relationship.

The evaluation begins with a simple assessment of numbers. Are there many more adverse event cases when the intervention is present, and fewer disease cases when the intervention is absent? If this question has been affirmatively answered, other questions follow. Does a greater exposure to the risk factor produce a greater extent of disease? Other questions asked by Hill explore the "believability" of the relationship. Some of these are: is there a discernible mechanism by which the risk factor produces the disease? Have other researchers also shown this relationship? Are there other such relationships whose demonstration helps us to understand the current risk factor– disease relationship?

Consistency requires that the findings of one study be replicated in other studies. The persuasive argument for causality is much more clearly built on a collection of studies involving different patients and different protocols, each of which identifies the same relationship between exposure to the risk factor and its consequent effect. There are numerous examples of collections of studies with different designs and patient populations, but that nevertheless successfully identify the same hazardous relationship between an exposure and disease. Identification of case series involving different series of patients in different countries and different cultures—yet each series producing the disease after the exposure would satisfy these criteria. Because research findings become more convincing when they are replicated in different populations, different studies that examine the same exposure–disease relationship and find similar results add to the weight of causal inference.

The specificity of a disease is directly related to the number of known causes of the disease. The greater the number of causes of a disease, the more nonspecific the disease is, and the more difficult it is to demonstrate a new causal agent is involved in the production of the disease. The presence of specificity is considered supportive but not necessary, and researchers no longer require that the effect of exposure to an agent such as a drug be specific for a single disease.

Exposure must occur before the disease develops for it to cause that disease. A temporal relationship must exist in order to convincingly demonstrate causation.* This criterion can be clearly satisfied by a case report that accurately documents that the exposure occurred before the disease.

* Protopathic bias, or the result of drawing a conclusion about causation when the disease process precedes the risk factor in occurrence can result without appropriate attention to the condition.

An evaluation of a relationship between occurrences of the adverse events and either the dose or duration of the intervention is also useful. The observation that a more intense exposure produces a greater frequency or severity of adverse events adds new strength to the notion that the relationship between the intervention and the adverse event is a causal one. In addition there should be some basis in the scientific theory that supports the relationship between the supposed "cause" and the effect. However, observations have been made in epidemiological studies that were not considered biologically plausible at the time but subsequently were shown to be correct.

It is important to note in the application of these tenets that satisfaction of all nine is not required to establish to the satisfaction of the medical community that a causative relationship exists between the exposure and the disease. Hill himself stated [3]:

> None of my nine viewpoints can bring indisputable evidence for or against the cause–and–effect hypothesis, and none can be required as a *sine qua non*.

However, these tenets are invaluable in the assessment of intervention-adverse event relationships.

7.5.3 The Need for Balance

Because early data suggesting a treatment effect may be misleading, investigators require additional support before they feel justified in terminating a research effort. Although an examination of the treatment effect over the course of the study may suggest that the research should end prematurely, other considerations may require the study to proceed to the end.

An example of this is the THRIVE III study [4]. Venous thromboembolic (VTE) phenomena (e.g., pulmonary emboli) are life-threatening events. Should the patient survive a first VTE, he is at risk for developing a subsequent VTE episode. The standard preventive therapy for the occurrence of additional VTE's is the administration of anticoagulant therapy for six months. However, whether an additional course of thromboembolic therapy beyond six months is necessary remains an open question. THRIVE III was a clinical trial that was prospectively designed to address this issue. Its objective was to assess whether use of oral coagulation prophylaxis after a six-month anticoagulation treatment for VTE reduced the recurrent rate of VTE.

THRIVE III was a double-blind, randomized, placebo-controlled parallel group multinational study that compared the efficacy of ximelagatran, a new oral anticoagulant, to placebo therapy. It recruited patients from Europe, South America, Canada, Mexico, Israel, and South Africa. There were 2466 patients randomized, 1223 of whom received the study drug. The prospectively declared primary analysis was the effect of therapy on recurrent VTE.

The findings for the study were striking. The placebo group experienced a 12.6% risk of recurrent VTE over the 18-month follow-up. The risk for recurrent VTE in the treatment group was 2.8%, substantially less than that seen in the control group arm of the study. The risk ratio was 0.16 (95% CI 0.09 to 0.30, p = 0.0001). Furthermore, the graphical depiction of the THRIVE III results reveals a clear early separation of the cumulative incidence of the primary endpoint (Figure 4.1).

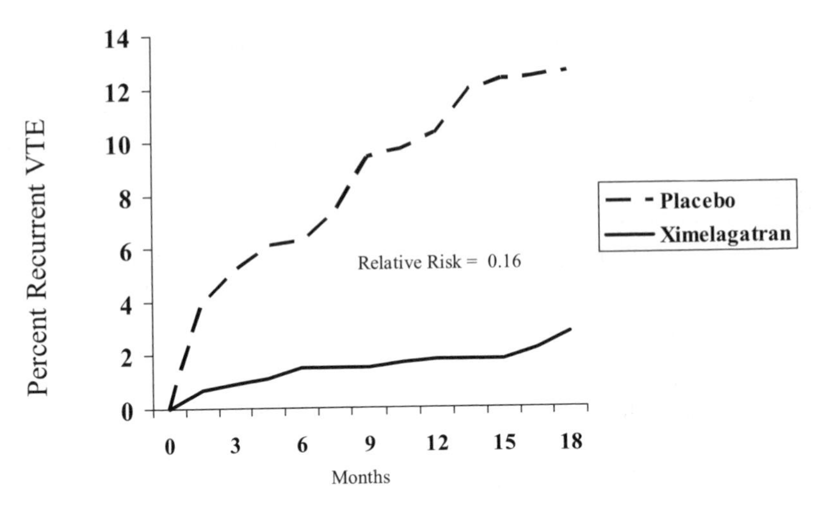

Figure 7.5. Percent reduction in recurrent venous thromboembolism (VTE) for ximelagatran versus placebo in THRIVE III, demonstrating an early treatment benefit.

The separation between these curves began immediately and continued inexorably for 18 months, demonstrating a profound treatment effect over the course of the study. However, the clear early emergence of a treatment effect begs the question of why was it necessary to continue this study for 18 months. Could the study not have been stopped at 15 months, or even 12 months of follow-up?

7.5.4 The Clinical Ethic versus Mathematics

The early and accelerating separation between the event curves of the placebo and active group event rates suggests that early termination was possible. In fact, an interim evaluation of the THRIVE III results supported this conclusion. However, the situation for the DMC was more complex. A major adverse effect of anticoagulation is bleeding. The risk of a major bleeding event, not depicted in Figure 4.1, required additional time to elucidate. Also, evidence of liver toxicity developed, an adverse event whose precise incidence estimate required additional follow-up time. Thus, the emergence of an early treatment effect, although suggesting that early termination may be necessary, does not provide the entire picture. An important motivation for continuing the trial in the face of this clear clinical benefit is the identification of other effects of the drug.

7.6 Monitoring for Futility

It is ethically important to monitor a research effort for premature evidence of benefit or harm to the participants. However, early termination of the study in the face of an impending null finding (i.e., a finding of neither benefit nor harm) has important advantages as well.

Prematurely concluding a clinical research effort because it is likely that the test statistic will not fall in the critical region, and will therefore provide a null finding, has become known as stopping for *futility* in clinical research jargon. The use of the term is understandable, but unfortunate. Investigators have powerful motivations to conduct research that they believe provides a palpable benefit to their patients. When it is unlikely that their research effort will provide this finding, and will thus contradict their hopes, it is easy to think of the research exercise as futile.

However, well-designed clinical studies are informative regardless of whether the findings are beneficial, harmful, or null. The identification of a null finding can and should close down a promising but ultimately unproductive avenue of research. This premature termination will allow the allocation of unspent human and financial resources to other investigative possibilities. The quicker the result can be determined, the sooner this reallocation can commence. Resources for clinical trials are finite and, in, an increasingly cost-conscious environment in which many promising interventions compete for scant research dollars, the earlier the study can be reliably discontinued, the better [5].

This observation has resonated in the lay press. The *New York Times* [6] reported that only 8% of pharmaceutical products that are studied ever make it to market. Even a small improvement in the ability to predict an early outcome, particularly early failures, could save hundreds of millions of dollars in drug development costs.

We might expect, that, just as the probability tools that we have described thus far can produce effective measures of the early emergence of efficacy or safety, they can also produce measures of how likely the study will produce neither. A simple adaptation of the use of conditional power can provide very useful and comprehensible measures of "futility".

7.6.1 Constructing a Futility Index

The development of the statistical argument for futility using conditional power is analogous to that for efficacy. In that case, the investigator was willing to consider early termination under a collection of conservative circumstances. In the process of monitoring for efficacy, the finding that the test statistic was likely to fall into the critical region was not sufficient to discontinue the study. This likelihood had to be large under the circumstance that no subsequent efficacy was produced by the intervention. If, based on the eventuality that the only effectiveness of the intervention had already been seen, and no new efficacy would occur, it was still likely that the test statistic would fall in the critical region at the study's conclusion, then the investigator could consider terminating the study early. This is a very conservative and useful condition.

This style of conservative thinking can be readily applied to the condition of early termination for lack of effect. Here, the efficacy of the intervention is absent up to an information time point I in the study (Figure 7.6).

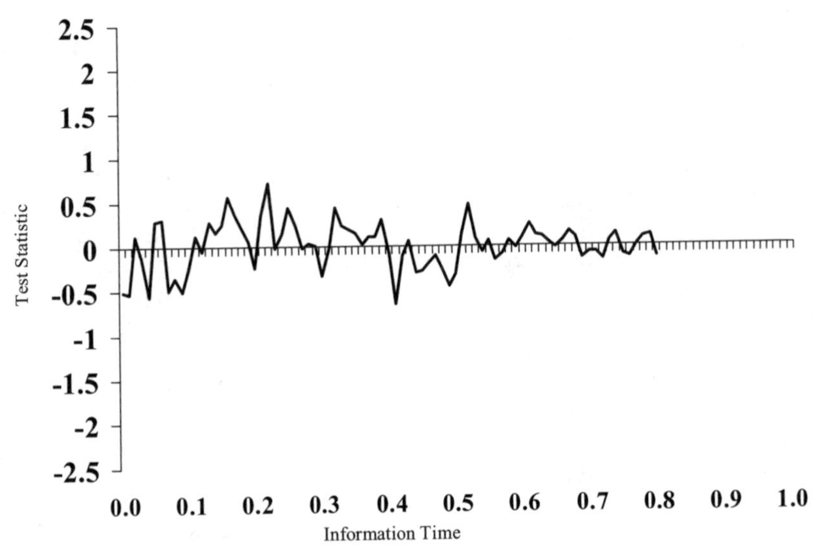

Figure 7.6. Possible values of a test statistic versus information time. Note the complete absence of a trend, suggesting that the study may be terminated early.

In Figure 7.6, the clinical research effort has proceeded until 80% of the information time has elapsed, with no sign of effectiveness of the studied intervention. One approach to the consideration of futility would be to just consider stopping the trial if this set of circumstances were to continue to occur. However, a more conservative tack to take would be to contemplate early termination even if some degree of late efficacy occurred. Specifically, the trial might be stopped even when new, late efficacy reveals itself, but this efficacy was not sufficient to lift the test statistic into the critical region of the study (Figure 7.7).

In Figure 7.7, the study results have revealed no positive or useful measure of effectiveness. If the study were to continue with the same lack of effect, an argument could be made for early futility. However, even with the presence of late efficacy, it may still be too late for the findings of the study to reverse themselves and produce a test statistic that will fall in the critical region at the conclusion of the study. Figure 7.6 shows several different, late paths of the test statistic that changes the direction for the test statistic. Each one of these paths assumes a different value of the efficacy that itself appears late in the study. However, for each of these paths, the test statistic still fails to fall in the critical region, and therefore, produces a null result.

Therefore, the conservative solution for futility is one that assumes that late efficacy appears in the study. If, even with the occurrence of this late efficacy the test statistic is not likely to fall in the critical region, researchers can consider terminating the research effort early because of lack of effect.

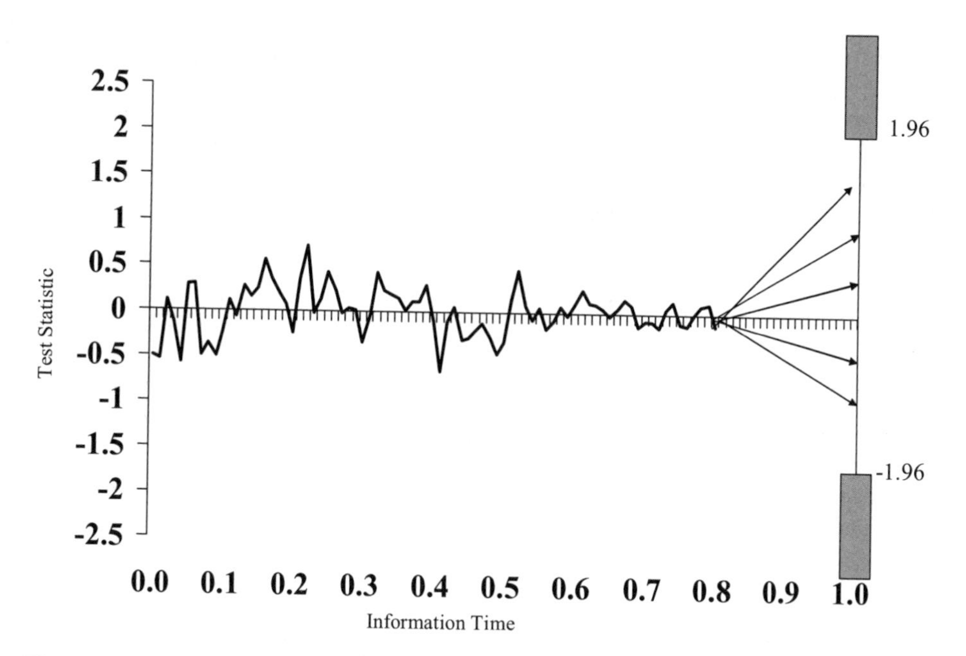

Figure 7.7. Trajectory of a futile outcome. The arrows represent the late efficacy (or harm) that could appear but still be associated with a null outcome of the study.

7.6.2 Calculation for Futility

Our plan is to identify an event that captures the dynamics of the futile research circumstance, including the appearance of late efficacy. Once we have that event, we can use what we know about conditional power to find its probability.

We seek to identify a boundary value f for the test statistic that has a value at a particular information time I, $TS(I)$. This boundary will determine whether consideration should be given to ending the trial. If the value of the test statistic at information time I falls below the boundary value f, then it is very unlikely that the test statistic will fall in the critical region at the conclusion of the study, and consideration should be given to ending the trial for futility. Alternatively, if $TS(I)$ is greater than f, then the study should be continued because there is substantial probability that the study will be positive.

Furthermore, the lower the probability that the test statistic falls in the critical region at the conclusion of the trial, the more futile the result. Thus, high "futility" is linked to small values of conditional power. This is unlike the circumstance that led to consideration of study termination for either efficacy or safety. In those situations, high value for conditional power leads to early termination. In the case for futility, the trial will be terminated in the event of low conditional power.

In the conditional power environment, this may be written as

$$CP_f = \upsilon = \boldsymbol{P}\left[TS(1) \geq Z_{1-\alpha/2} \mid TS(I) = f\right].$$

We subscript the conditional power with f to denote we are computing conditional power computation for futility. In this circumstance, we allow v to be a small value, that is, $v = 0.10$. Our goal is to identify the value of the test statistic f at information time I such that, any value of $TS(I)$ that is less than f suggests that the study should be terminated early. The development in Appendix C carries out the mathematics of this calculation in detail, and we find that the value of f is

$$f = \frac{Z_{1-\alpha/2} - Z_{1-\upsilon}\sqrt{1-I} - \mu(1-I)}{\sqrt{I}}.$$

This solution is a function of several quantities that we have come to recognize. The quantity $Z_{1-\alpha/2}$ denotes both the type I error rate at which the statistical significance of the test will be assessed at the conclusion of the study, and the sidedness of the test.* As anticipated, f is also a function of the conditional power v which is mandated to be low, and the information time I when the futility assessment is to be made. However, there is a new component that is included in this computation, μ. This is the drift parameter, associated with the alternative hypothesis.

7.6.3 Futility Under Alternative Drift*

Recall in Chapter Six that conditional power can be computed under not just one, but a collection of assumptions about the direction of the test statistic. When considering predictions about the location of the test statistic at the conclusion of the trial based on its value at information time I, we determined that the conservative solution would be to assume that all of the efficacy that had occurred in the trial occurred prior to the current information time I. This was equivalent to believing that there would be neither positive nor negative drift of the test statistic for the remainder of the study, that is, the drift parameter μ for the remainder of the study was 0. This we denoted as $C_p(H_0)$.

As pointed out earlier, the situation is different when faced with the possibility of early termination for futility. True futility occurs when, even in the face of late efficacy, the study is not likely to arrive at a positive conclusion. Thus, we must assume that there is upward drift for the $1-I_1$ unexpired duration of the study. If the test statistic is still not very likely to enter the critical region at the conclusion of the study, the study might very well be stopped at information time I_1 because it is unlikely to be positive.

In a clinical study designed to compare the relative frequency of events between two treatment groups (i.e., mortality), the drift parameter is

$$\mu = -\ln(R)\sqrt{E/4},$$

* One-sided or two-sided.

where R is the relative risk associated with the therapy effect and E is the total number of events that are expected in the study. For relative risks that reflect a risk reduction, the drift parameter is positive (Table 7.7).

Table 7.7. Drift Parameters in a Clinical Trial Designed to Determine Efficacy as a Funtion of the Relative Rrisk and Total Number of Endpoints

		Relative Risk					
		0.65	0.70	0.75	0.80	0.85	0.90
	50	1.52	1.26	1.02	0.79	0.57	0.37
	100	2.15	1.78	1.44	1.12	0.81	0.53
Total	200	3.05	2.52	2.03	1.58	1.15	0.75
endpoints	300	3.73	3.09	2.49	1.93	1.41	0.91
	400	4.31	3.57	2.88	2.23	1.63	1.05
	500	4.82	3.99	3.22	2.49	1.82	1.18
	600	5.28	4.37	3.52	2.73	1.99	1.29

In the scenario of a clinical research effort designed to measure the occurrence of discrete events, we may write

$$f = \frac{Z_{1-\alpha/2} - Z_{1-\upsilon}\sqrt{1-I} + \ln(R)\sqrt{E/4}(1-I)}{\sqrt{I}}. \tag{7.3}$$

The determination of the relative risk R and E which are used in the drift calculation that is necessary for futility is the subject of discussion by investigators and not uncommonly, by the DMC. Some argue that the best measure of these two quantities comes from the study protocol, where these quantities are specified and used in the study sample size determination. Others suggest that the relative risk and projected total number of events should come from the estimates generated by the study itself. However, each of these can easily be presented to the DMC.

The following is an illustration of how efficacy/safety boundaries can be developed for a clinical trial. Consider a study that is designed to assess the effect of a novel ultrasound therapy on the occurrence of the combined endpoint of fatal and nonfatal stroke rate in patients at risk for stroke. A major risk factor for stroke is the development of left atrial enlargement. The enlarged chamber can lead to the generation of blood clots, that, when broken free from the main clot on the atrial wall, travel through the left side of the heart to the organ systems of the body. Sometimes these dislodged clots travel to the brain where they can produce a stroke. The intervention to be studied in this trial is a unique hypersonic device that is designed to break up these atrial blood clots. When appropriately focused on the atria, the delivered ultrasonic will break the clot up, reducing the risk of embolic stroke.

All patients must have left atrial enlargement when they enter the study. Subjects who meet these criteria are recruited into one of two treatment groups. The control treatment group receives standard preventive therapy for stroke. The active group receives, in addition to this therapy, regular ultrasonic debridement of the left atrium.

This ambitious study is anticipated to take six years to complete. The expected cumulative fatal/nonfatal stroke rate in the control group is 13%. The investigators believe that this novel therapy will reduce the fatal/nonfatal stroke rate by 25%. They would like to carry out the statistical hypothesis test on the primary endpoint at 95% power, with an alpha error rate of 0.0225 (for efficacy only; type I error level for harm is considered separately below), after adjustment for 90% conditional power.* These considerations produce a sample size of 5074 patients, with 2537 subjects recruited to each of the two groups. The investigators plan to monitor the trial using 90% conditional power for efficacy.

As they plan the trial, the scientists initially consider monitoring at 10%, 30%, 50%, and 75% information time (Table 7.8).

However, the researchers are particularly nervous about the occurrence of harmful effects in the study associated with the intervention. This concern focuses on the possibility the ultrasonic treatment may produce thrombus flecks and debris that are insufficiently granulated by the process. Thus, it is feared that these flecks, once broken free of the atrial wall by the ultrasound, may travel through the outflow tract, into the carotid arteries, and to the brain. Once there, they can occlude a vessel, thereby producing a stroke. The investigators decide to allocate a type I error of 0.05 specifically to assess this possibility of harm. Thus, the total, prospectively declared type I error for the study is 0.0225 + 0.05 = 0.0725. To further amplify their concern for the occurrence of this dangerous event, the researchers choose to monitor safety at 80% conditional power.

---

* This adjustment is discussed in Chapter Six. Its purpose is to adjust the type I error rate for the multiple evolutions that take place during the monitoring process.

Table 7.8 Asymmetric Boundary Values for Early Termination (Two-Sided, Type I Error Rate = 0.0225, 95% Conditional Power for Benefit Type I Error Rate 0.050, 80% Conditional Power for Harm).

Information Time	Upper Boundary	Lower Boundary
0.10	11.27	-7.73
0.30	6.17	-4.29
0.50	4.48	-3.17
0.75	3.26	-2.39
1.00	2.00	-1.64

An additional concern of the investigators is that the intervention will be ineffective. The ultrasonic device has a panoply of nonserious adverse events associated with its use, and treatments must be repeated. In addition, it is expensive. The investigators would like to move on quickly to other possible interventions if the atrial hypersonic treatment is found to be ineffective. They therefore build in boundaries for the absence of effect into their monitoring rules.

Using equation (7.3) they can compute values of the test statistic below which there is inadequate demonstration of efficacy (Table 7.9).

Table 7.9 Boundary Values for Test Statistic for Futility as a Function of Information Time and the Conditional Power.

Information	Conditional power (v)		
time	0.10	0.15	0.20
0.50	-0.51	-0.34	-0.20
0.60	0.14	0.27	0.37
0.70	0.70	0.79	0.86
0.80	1.18	1.24	1.28
0.90	1.61	1.64	1.66
0.95	1.81	1.83	1.84
0.99	1.97	1.97	1.97

Table 7.9 provides the upper bound of the test statistic (i.e., the futility indices) as a function of the information time in the study. Three different conditional power assessments are examined: 0.10, 0.15, and 0.20. From this table, one may easily observe the relationship between the futility index and the conditional power as a function of information time (Figure 7.8).

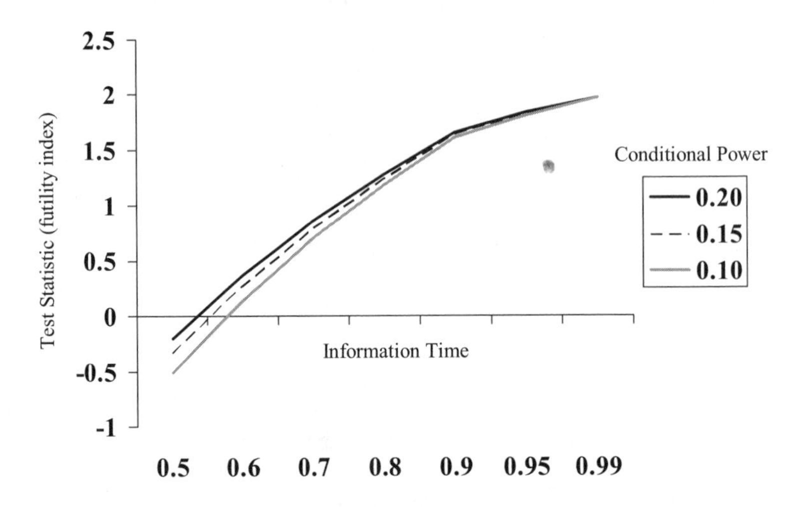

Figure 7.8. Test statistic (futility index) as a function of information time and conditional power.

Figure 7.8 requires careful consideration. Note that, for fixed information time, larger conditional powers are associated with greater boundary values, consistent with our findings for efficacy and safety.

However, Figure 7.8 demonstrates that the futility index or test statistic boundary increases as the information time increases. This is different from the relationship between efficacy (or safety) boundary values and information time, where we observed that the boundary values get closer to zero as the information time increases. This new relationship requires a closer examination.

At the relatively early information time of 0.50, the study really cannot be described as futile because, essentially, any positive value of the test statistic is above the futility boundary, a finding consistent with continuing the trial. However, as follow-up time in the study increases, the effect of future efficacy diminishes, and there emerge positive values of the test statistic for which the study should be terminated early. This is because, even though the test statistic is positive, it is unlikely that, even with the full force of efficacy appearing for the remainder of the study, the test statistic will be lifted into the critical region at the study's conclusion.

Thus, the closer the study is to its planned termination, the smaller the influence of efficacy in the unelapsed information time, and the larger the test statistic may be without being moved into the critical region by the declining influence of future efficacy. The prospectively declared monitoring values for efficacy, safety, and futility can be combined into one figure (Figure 7.9).

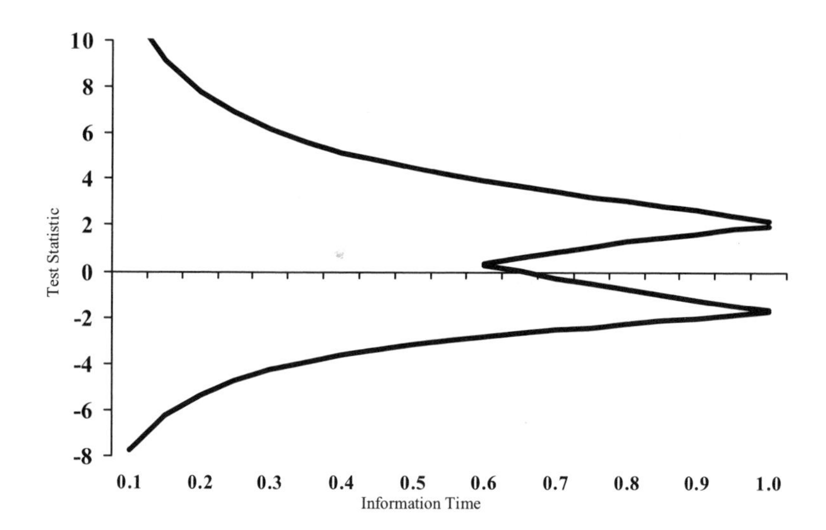

Figure 7.9. Monitoring guidelines for safety, efficacy, and futility in a clinical trial: alpha (benefit) = 0.0225, alpha (harm) = 0.05), Cp (benefit) = 0.95, Cp (harm) = 0.80, futility index (benefit) = 0.20, futility index (harm) = 0.10.

Figure 7.9, reveals the boundary values and decisions that are considered within each boundary crossed. Crossing the upper or lower boundaries leads to the consideration of terminating the study for early efficacy or harm. Crossing the boundary to the right leads to the contemplation of early termination for futility. The lower bound for futility is computed following the reasoning of the calculation of the upper futility boundary. The formula for this upper boundary is provided in Appendix C as

$$f = \frac{Z_{\alpha/2} - Z_{\upsilon_s}\sqrt{1-I_1} - \mu(1-I_1)}{\sqrt{I_1}},$$

where $\mu = -\ln(R)\sqrt{E/4}$. In this circumstance of futility for safety, the relative risk R is greater than one, signifying hazard.

It is important for the investigators to keep in mind that they control the contours of these regions by choosing the type I error, the conditional power, and the futility indices for both efficacy and harm. For example, if they envision stopping the trial for only extreme magnitudes of efficacy early, they can set the type I error for benefit to 0.001, 99% conditional power, and futility of 0.10. By setting alpha for harm at a high value, 0.10, in concert with a relatively low conditional power for harm (0.10), and low futility index 0.005, they are simultaneously quite sensitive to the early occurrence of harm (Figure 7.10).

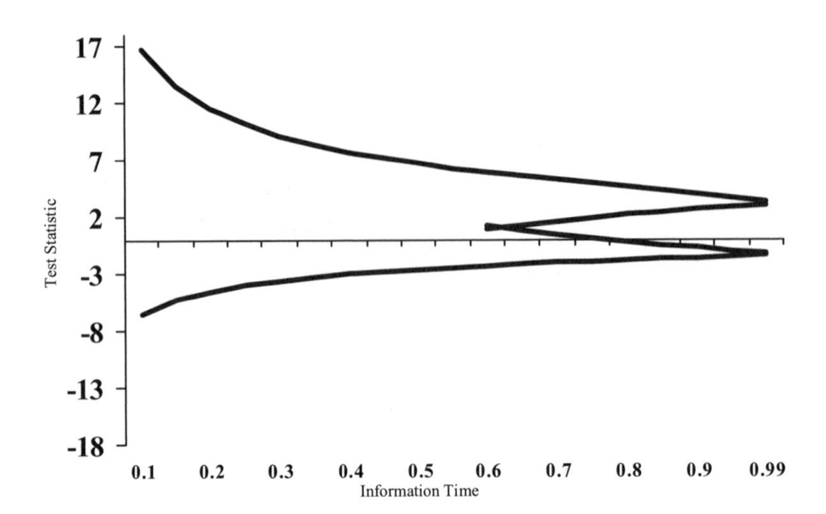

Figure 7.10. Alternative example for monitoring for safety, efficacy, and futility in a clinical trial. Alpha (benefit) = 0.001, alpha (harm) = 0.10), Cp (benefit) = 0.99, Cp (harm) = 0.80, futility index (benefit) = 0.10, futility index (harm) = 0.20.

Thus, investigators with their appreciation of the benefits and risks of the intervention can calibrate the monitoring process of the clinical research so that I provides the best patient-protective characteristics.

Problems

1. Describe the complications in interpreting exploratory safety evaluations versus exploratory efficacy evaluations.

2. An investigator is interested in monitoring a clinical research effort for efficacy and safety. The information time points at which the study will be monitored are $I = 0.40$ and $I = 0.80$. Compute the boundary values for 90% conditional power for the test statistic assuming a two-sided alpha error of 0.05.

3. An investigator is interested in monitoring a clinical research effort for efficacy and safety. The information time points at which the study will be monitored are $I = 0.40$ and $I = 0.80$. Compute the boundary values for 90% conditional power for the test statistic assuming an alpha error of benefit of 0.01, and an alpha error for harm of 0.04.
4. An investigator is interested in monitoring a clinical research effort for efficacy and safety. The information time points at which the study will be monitored are $I = 0.40$ and $I = 0.80$. Compute the boundary values for 90% conditional

power for benefit and 80% conditional power for harm assuming an alpha error of benefit of 0.01, and an alpha error for harm of 0.04.

5. What are the advantages of the use of a combined adverse event endpoint?

6. An investigator is carrying out a small prospective clinical study to assess the effect of a promising new agent to reduce the severity of ischemic bowel syndrome. A serious adverse effect of this condition is gastrointestinal obstruction. In this ill cohort of patients, it is anticipated that 16% will obstruct. Currently, the investigator has treated 30 patients in her study and has observed 3 patients who developed gastrointestinal obstruction after use of the new active agent. If another 40 patients are to be recruited into the research effort, what is the minimum number of patients that will suffer obstruction in this new cohort order so that the 95% confidence interval for the obstruction rate centered on a rate of 0.16 is exceeded.

7. In what circumstances will use of the Hill causality tenets adumbrate the reliance on statistical tools in monitoring clinical research?

8. Why does the development of the conservative futility concept assume that there is underlying efficacy in the remainder of the study, but the underlying conservative conditional power for efficacy computation assumes that there is no efficacy for the unexpired duration of the study?

9. A clinical trial recruits 600 patients in order to assess the efficacy of a therapy for toxic megacolon. The background mortality rate is 40%, and the investigators anticipate that they will produce a 25% reduction in this rate. The two-sided type I alpha rate is 0.05 and 80% power. Compute the futility index for efficacy at information times of 70%, 80%, and 90% for a conditional power of 10%.

References

1. Moyé LA. (2003) *Multiple Analyzes in Clinical Trials. Fundamentals for Investigators*. New York, Springer. Chapters 7-8.
2. Wald A. (1947) *Sequential Analysis* New York, John Wiley and Sons
3. Hill B. (1953) Observation and experiment. *New England Journal of Medicine* **248**:995–1001.
4. NDA 21-686. FDA Advisory Committee Briefing Document. September, 2004.
5. Ware JH, Muller JE, Braunwald E. (1985) The futility index. An approach to the cost-effective termination of randomized clinical trials. *American Journal of Medicine* **78**:635-643.

6. Pollack A. In drug research, the guinea pigs of choice are now, well, human *The New York Times*, August 4, 2004.

8

Bayesian Statistical Monitoring

Scientific progress ensures that the innovative ultimately become commonplace. Such is the case with monitoring procedure in clinical trials. The exciting work in the 1970s and 80s that produced the group sequential procedures of Pocock [1], O'Brien–Fleming [2], Lan–DeMets [3,4], the triangle procedures of Whitehead [5,6], and the conditional power approach of Halperin [7], have become the accepted norm in clinical trial research, and their implementation has standardized the interpretation of interim results of these studies.

Bayes procedures are important new innovations in the statistical monitoring of clinical trials. Their emphasis on what has come to be known as the "likelihood principle", or the concept that one need focus only on what has occurred, rather than what has not happened, provides a useful, exciting, and sometimes illuminating perspective on the clinical research paradigm. However, Bayes procedures do not come without their own risks. The contribution of "prior information" and the construction of a "loss function" can be obstacles to both the construction of the Bayes result and the acceptance of that result by the medical community. This chapter focuses on the incorporation of the Bayes philosophy into the design of clinical research.

8.1 Frequentist versus Bayesian Philosophy

The frequentist philosophy is the cornerstone of classical mathematical statistics and hypothesis testing. Based on probability, it focuses on the relative frequency of the occurrence of an event.

However, there are two concerns about this approach that disturb some workers. The first is its reliance on what has not happened, a consequence of the definition of relative frequency. The second criticism is the frequentist philosophy's silence on the reliability of one experiment. The frequentist perspective concentrates on long-term, rather than short-term reliability.

As a simple illustration, consider the experiment of flipping three fair coins simultaneously, and asking how likely it is each will show a head. In computing this probability, we simply count the eight different events {HHH}, {HTH},

{HTT}, {HTH}, {THH}, {THT}, {TTT}, and {TTH}. Because each is equally likely, we compute that the probability of {HHH} is 1/8. Note that the computation of this probability focuses on not only what was observed, but on the seven possible events that did not occur. Other examples of the difficulty considering what has not occurred are discussed by Lindley [8] and Berger [9].

Secondly, stating that the event {HHH} will occur 1 in 8 times is not meant as an exact prediction to be applied to all flips of the three coins. We cannot guarantee that if the three coins are flipped 8 times, {HHH} will appear once, or that in 16 flips, {HHH} will appear twice. The solution 1/8 refers to the long run frequency of occurrence. If one were to carry out the experiment 100 times, or better yet, 1000 times, we would expect the sequence {HHH} to occur in $1/8^{\text{th}}$ of the flips. We computed the estimate 1/8 not as a measure of short-term behavior, but as a long-term predictive measure.

While classical statisticians develop procedures and estimates that are accurate in the long-term, they are relatively mute on the interpretation of an estimator from a single experiment, stating instead that, if the experiment were repeated, the sample estimates would be close to the population values many more times than not. However, as we pointed out in Chapter Two, the researcher's perspective is not focused on long-term accuracy. The scientists, after having expended a good deal of their own work (and, commonly, other's people's money) in the research effort, are focused on how accurate their single experiment is.

The interpretation of the *p*-value serves as a fine example of this dilemma. Suppose an investigator conducting a clinical trial in patients susceptible to stroke, assesses the effect of an active therapy versus control therapy on the observed fatal stroke rate. The *p*-value generated from this evaluation is the probability that a population in which the therapy is not effective produces a sample that demonstrates the observed treatment effect. However, what the researcher really needs to know is whether his particular results were produced by simple chance and sampling error* or instead were driven by a true therapy effect in the population. He has a question about his specific result; yet this is a question that is not answered by the *p*-value, which instead focuses on what would occur if many experiments were carried out.

Like classical statistics, Bayes philosophy is applicable to problems of parameter estimation and hypothesis testing. However, there are several important differences between the Bayesian and frequentist approaches. One is that the Bayesian formulation is based on the *likelihood principle.* The likelihood principle states that a decision should have its foundation in what has occurred, not in what has not happened.

In addition, although Bayesians, like frequentists, are interested in parameter estimation and hypothesis testing. Bayesians do not believe that the parameter θ of a distribution is constant, but instead believe that the parameter has a probability distribution. This is called the *prior distribution,* or $\pi(\theta)$. For example, when at-

* Sampling error is the concept that a population will produce different samples, and that these samples, containing different subjects with different life experiences, will contain different estimates of the same effect. This variability from one sample to another is termed sampling error. This is discussed in Chapter Two.

tempting to identify the mean change in blood glucose for a collection of individuals, both the frequentist and the Bayesian may assume that the distribution of blood glucose for this sample of individuals follows a normal distribution with an unknown mean, whose estimation is the goal. However, the frequentist treats this unknown mean as a fixed parameter. The Bayesian assumes that the parameter has its own probability distribution.

Once the prior distribution is identified, the Bayesian works forward, next identifying the probability distribution of the data given the value of the parameter. This distribution is described as the *conditional distribution* (because it is the distribution of the data conditional on the value of the unknown parameter) and is denoted as $f\left(x_1, x_2, x_3, \ldots, x_n | \theta\right)$. This step is not unlike that of the frequentist. The Bayes process continues by combining the prior distribution with this conditional distribution to create a *posterior distribution*, or the distribution of the parameter θ given the observed sample, denoted as $\pi\left(\theta | x_1, x_2, x_3, \ldots x_n\right)$.

From the Bayes perspective, the prior distribution reflects knowledge about the location and behavior of θ before the experiment is carried out. The execution of the experiment provides new information that is combined with the prior information to obtain a new estimate of θ. To help in interpreting the posterior distribution, some Bayesians will construct a loss function that identifies the penalty that they pay for underestimating or overestimating the population parameter. Bayesian hypothesis testing is based on the posterior distribution,

As we pointed out in Chapter Two, the Bayesian approach to statistical analysis makes unique contributions. It explicitly considers prior distribution information, providing direct, new, and important input into the computational procedure. It allows construction of a loss function that directly and clearly states the loss (or gain) for each decision. Bayes procedures have admirable flexibility in cases where the relationship between the frequentist's p-value is disconnected from the loss the community suffers in the face of a mistaken scientific conclusion. However, the requirement of a realistic specification of the prior distribution can be a burden if there is no information about the parameter to be estimated. Similarly, the choice of the loss function can be difficult to justify from a clinical perspective.

The work of Spiegelhalter et al. [10] lays out the basics of the application of the Bayes concept to monitoring rules in clinical research. This work uses a conditional power argument as part of its methodology, a procedure with which we are comfortable. Throughout the following illustrations of the Bayes perspective on clinical trial monitoring, we will use this approach, pointing out the distinctions between the Bayesian and frequentist philosophies.

8.2 An Example of a Bayes Monitoring Rule

In this first illustration, we focus on an investigator who wants to determine if the cranial application of ultrasound reduces the progression of stroke. She focuses on patients admitted to her stroke service, planning to randomize patients who have suffered a stroke to either the active group or the control group. To blind the study, every patient receives a "sonic treatment". For the patients in the control group, this consists of merely mild and harmless buzzing, meant to simulate the sensation of ultrasound. Patients in the active group receive the actual ultrasound treatment.

All patients will be followed until they are discharged from the hospital. The endpoint of the study is the duration of hospital stay, which she has observed has a mean of 13 days and standard deviation of 4 days. The researcher hopes to demonstrate that patients who receive the ultrasound will have a hospital stay that is on average three days shorter then the average hospital stay of patients in the control group. She will carry out the hypothesis test at the end of the trial with a two sided type I error rate of 0.05 and a power of 90%. These assumptions produce a sample size of 76 patients, 38 in each group. She will monitor the clinical study at $I = 0.50$, that is, when 50% of her randomized cohort have been discharged from the hospital.

This information is typically enough for the frequentist statistician to plan the evaluation of both the data at the monitoring time point and the final assessment of the data at the conclusion of the study. However, the traditional Bayesian requires there to be a more complete elaboration of one concept.

8.2.1 Prior Distribution

The additional step required by the Bayesian is the probability distribution of the effect produced by the ultrasound. If we define this difference in days hospitalized between the active and the control group as θ, then the investigator has stated the problem as though there were only two possible values for this parameter: $\theta = 0$ consistent with the null hypothesis of no reduction in days hospitalized, or $\theta = -3$, consistent with an ultrasound-induced three day reduction. However, the Bayesian permits greater freedom of choice for the reduction in hospital days through the specification of a probability distribution on the possible values of θ. This will be $\pi(\theta)$.

The prior distribution can be any known probability distribution that is appropriate for the setting.* It is up to the investigator to specify the probability distribution function. As its name implies, it is best if based on "prior information" from the medical or research community. In this example, the investigator chooses to let θ be any one of –5, –4, –3, –2, –1, or 0. We can assign probability to each of these values, defining the prior distribution for θ (Table 8.1).

* Sometimes, Bayesians will actually use a distribution for $\pi(\theta)$ which is not a probability distribution, that is, the sum of the probabilities for all disjoint events is not one. This is called an *improper prior*, and leads to important interpretative complexities.

Table 8.1. Probability Distribution for Effect of Therapy Designed to Reduce Post-Stroke Hospitalizalion Duration

Reduction	Probability
0	0.05
-1	0.15
-2	0.20
-3	0.30
-4	0.20
-5	0.10

We see from Table 8.1 that the investigators believe that there is a 15% chance that the actual reduction in post-stroke hospital duration is 1 day. Mathematically, we would write this as $\pi(\theta) = 0.15\mathbf{1}_{\{\theta=-1\}}$, where the quantity $\mathbf{1}_{\{\theta=-1\}}$ is equal to when $\theta = -1$ and 0 otherwise. Because there are other values of θ, each of which have their own probability, we proceed by writing the entire prior distribution as

$$\pi(\theta) = 0.05\mathbf{1}_{\{\theta=0\}} + 0.15\mathbf{1}_{\{\theta=-1\}} + 0.20\mathbf{1}_{\{\theta=-2\}} + 0.30\mathbf{1}_{\{\theta=-3\}} + 0.20\mathbf{1}_{\{\theta=-4\}} + 0.10\mathbf{1}_{\{\theta=-5\}}. \quad (8.1)$$

which simply reflects the six different possible values of θ. In general if there are k possible values for θ, θ_1, θ_2, θ_3, ... θ_k, with probabilities p_1, p_2, p_3, ..., p_k. then we would write the prior distribution $\pi(\theta)$ as $\pi(\theta) = \sum_{k=1}^{K} p_k \mathbf{1}_{\{\theta=\theta_k\}}$.

8.2.2 Computing the Power

The investigator now has a wider range for possible values of effectiveness, each attached to a prior probability. She can now consider the probability that the test statistic will fall in the critical region at the trial's conclusion given the value of the test statistic at information time I. Recall from Chapter Six that we wrote that the probability the test statistic at information time I_2, $TS(I_2)$ will be greater than value s_2 given that at an earlier time I_1 the test statistic $TS(I_1)$ is equal to some value s_1 as $\boldsymbol{P}\left[TS(I_2) \geq s_2 \mid TS(I_1) = s_1\right]$. We showed that

$$\boldsymbol{P}\left[TS(I_2) \geq s_2 \mid TS(I_1) = s_1\right] = \boldsymbol{P}\left[\sqrt{I_2}TS(I_2) \geq \sqrt{I_2}s_2 \mid \sqrt{I_1}TS(I_1) = \sqrt{I_1}s_1\right]$$
$$= \boldsymbol{P}\left[B(I_2) \geq \sqrt{I_2}s_2 \mid B(I_1) = \sqrt{I_1}s_1\right] = \boldsymbol{P}\left[B(I_2 - I_1) \geq \sqrt{I_2}s_2 - \sqrt{I_1}s_1\right].$$

In the case of our investigator, $I_1 = 0.50$, $I_2 = 1$, and $s_2 = 1.96$. She can write

$$\boldsymbol{P}\left[TS(1) \geq 1.96 | TS(0.50) = s_1\right] = \boldsymbol{P}\left[B(0.50) \geq 1.96 - s_1\sqrt{0.50}\right]. \tag{8.2}$$

If this were going to be the standard conditional power computation that we developed in Chapter Six, we would make the conservative assumption that the mean of the Brownian motion process was zero. This we saw was equivalent to assuming that there would be no additional efficacy in the unexpired duration of the study. This assumption would permit us to complete the computation in expression (8.2) to find

$$\begin{aligned}
\boldsymbol{P}\left[TS(1) \geq 1.96 | TS(0.50) = s_1\right] &= \boldsymbol{P}\left[B(0.50) \geq 1.96 - s_1\sqrt{0.50}\right] \\
&= \boldsymbol{P}\left[N(0.50,\ 0.50) \geq 1.96 - s_1\sqrt{0.50}\right] \\
&= \boldsymbol{P}\left[N(0,1) \geq \frac{1.96 - s_1\sqrt{0.50}}{\sqrt{0.50}}\right] \\
&= 1 - F_Z\left(2.77 - s_1\right).
\end{aligned} \tag{8.3}$$

However, because the Bayesian model acknowledges the availability of prior information, our task is to now incorporate it into the conditional power computation. Recall that, in the presence of drift, the calculation of expression (8.3) becomes

$$\begin{aligned}
\boldsymbol{P}\left[TS(1) \geq 1.96 | TS(0.50) = s_1\right] &= \boldsymbol{P}\left[B(0.50) \geq 1.96 - s_1\sqrt{0.50}\right] \\
&= \boldsymbol{P}\left[N(0.50\mu,\ 0.50) \geq 1.96 - s_1\sqrt{0.50}\right] \\
&= \boldsymbol{P}\left[N(0,1) \geq \frac{1.96 - s_1\sqrt{0.50} - 0.50\mu}{\sqrt{0.50}}\right] \\
&= 1 - F_Z\left(2.77 - s_1 - \sqrt{0.50}\mu\right),
\end{aligned} \tag{8.4}$$

where μ is the drift parameter of the Brownian process. The investigator's task is to now convert the prior information about the mean change in duration of hospital days θ into a statement about the drift parameter μ. This we can easily accomplish. Recall from Chapter Six that the drift parameter is

$$\mu = \sqrt{\frac{N}{2\sigma^2}}\theta, \tag{8.5}$$

where N is the half of the number of subjects required for the study and σ_Δ^2 is the variance of the change in hospital stay within the treatment group.*

---

* This assumes the number of subjects recruited to each group is the same. The demonstration follows easily from the observation that if x_i is the hospital stay of an active group pa-

Because the investigator knows that the prior distribution of θ is $\pi(\theta)=\sum_{k=1}^{K} p_k \mathbf{1}_{\{\theta=\theta_k\}}$. for $K = 5$, the prior distribution for the drift parameter μ is

$$\pi(\mu)=\sum_{k=1}^{K} p_k \mathbf{1}_{\left\{\mu=-\sqrt{\frac{N}{2\sigma^2}}\theta_k\right\}}.$$

This simply means that for every possible value of θ under the prior distribution, we take that value and multiply it by $\sqrt{N/2\sigma^2}$ to get the possible value of the drift parameter μ.

8.2.3 Incorporating the Prior Information

At this point, the investigator has developed the prior information for the effect of therapy in this research effort that was designed to demonstrate that a clinical intervention during the evolution of a stroke could reduce hospital duration. This prior information was transformed into the parameter θ, the difference in hospital stay between the intervention group and the control group.

We have also developed the conditional power for monitoring the study at 50% information time. This was a function of μ, the drift parameter. Finally, we linked μ to θ in such a way to have the prior distribution for μ. It now remains to incorporate μ into the conditional power development.

Recall that we need to compute the conditional power which we saw was

$$\boldsymbol{P}\left[TS(1)\geq 1.96 \mid TS(0.50)=s_1\right]=\boldsymbol{P}\left[N(0.50\mu,\ 0.50)\geq 1.96-s_1\sqrt{0.50}\right]$$

---

tient, and y_i is the duration of hospital stay of a control group patient, then $x_i - y_i$ follows a normal distribution with mean θ and variance $2\sigma^2$. From this we can see that $\dfrac{\sum_{i=1}^{n} x_i - \sum_{i=1}^{n} y_i}{\sqrt{2\sigma^2 N}}$ follows a normal distribution with mean

$$\frac{n}{N}\sqrt{\frac{N}{2\sigma^2}}\theta$$

and variance n/N. Because the information time $I = n/N$, we can write the mean of this distribution as

$$\frac{n}{N}\sqrt{\frac{N}{2\sigma^2}}\theta$$

or as μI where

$$\mu=\sqrt{\frac{N}{2\sigma^2}}\theta.$$

when μ itself has a probability distribution. We have the probability given the value of μ, which is simply

$$\begin{aligned} \boldsymbol{P}\left[TS(1) \geq 1.96 \mid TS(0.50) = s_1\right] &= \boldsymbol{P}\left[N(0.50\mu,\ 0.50) \geq 1.96 - s_1\sqrt{0.50}\right] \\ &= 1 - F_Z\left[\frac{1.96 - s_1\sqrt{0.50} - 0.50\mu}{\sqrt{0.50}}\right] \\ &= 1 - F_Z\left[\frac{1.96 - s_1\sqrt{0.50} - 0.50\sqrt{\frac{N}{2\sigma^2}}\theta}{\sqrt{0.50}}\right]. \end{aligned}$$

Using the observation that the prior distribution is simply a weighted average of the drift parameters, we may write

$$\boldsymbol{P}\left[TS(1) \geq 1.96 \mid TS(0.50) = s_1\right] = \sum_{k=1}^{k} p_k \left(1 - F_Z\left[\frac{1.96 - s_1\sqrt{0.50} - 0.50\sqrt{\frac{N}{2\sigma^2}}\theta}{\sqrt{0.50}}\right]\right)$$

the weighted average of the conditional power for each of the six different values of θ.

Assume that the investigator has observed a test statistic of 2.75 at the 50% information time point, she may compute the conditional power of her study (Table 8.2).

Table 8.2. Predictive Power for Test Statistic of 2.75 (I=0.50, Two-Sided Type I Error = 0.05)

θ	$\pi(\theta)$	Drift Parameter	Conditional Power
0.0	0.05	0.000	0.491
-1.0	0.15	1.541	0.857
-2.0	0.20	3.082	0.985
-3.0	0.30	4.623	0.999
-4.0	0.20	6.164	1.000
-5.0	0.10	7.706	1.000
		Predictive Power =	0.950

The final computation has become known as *predictive power*. Note that the computation is actually straightforward. The investigator simply computes the conditional power for each of the values of efficacy θ, and then averages them, using the values $\pi(\theta)$ as the weights. As we would expect, the larger the values of efficacy, the greater the permitted drift for the unexpired duration of the study, and the greater the conditional power.

As we might expect, the final predictive power is a function of the prior distribution $\pi(\theta)$. Alternative assumptions about the possible values of the effectiveness of the ultrasound treatment produce very different assumptions about the predictive power at $I = 0.50$ (Table 8.3). In this setting, we see that more conservative assumptions about the effect size of the ultrasound treatment's effect, which placed greater probability on smaller reductions in the hospitalization duration, produced a strikingly smaller predictive power.

It is important to note that each of the computations is correct only for the prior distribution on which it is based. Thus, the investigators are best served by a clear understanding of the prior distribution and its influential role in computing predictive power.

Table 8.3 Predictive Power for Test Statistic of 2.75, with an Alternative Assumption for the Prior Distribution π(θ). (I=0.50, Two Sided Type I Error = 0.05).

θ	π(θ)	Drift Parameter	Conditional Power
0.0	0.15	0.000	0.491
-0.5	0.35	0.771	0.700
-1.0	0.40	1.541	0.857
-2.0	0.05	3.082	0.985
-2.5	0.04	3.853	0.997
-5.0	0.01	7.706	1.000
		Predictive Power =	0.760

8.3 Using Continuous Prior Information

In the previous example, when the prior information was discrete, the Bayes predictive power was a weighted average of the conditional powers produced from each of the values of the therapy effectiveness.

However, prior information can be continuous as well as discrete. In these latter settings, the computations become somewhat more complicated, but, from the investigator's point of view, the solution is the same, that is, a weighted average of conditional powers.

As an example of the use of continuous prior information, continuous prior distribution for a clinical effect will be incorporated to create a Bayesian monitoring procedure for a clinical trial. In this experiment, patients who are suffering from an acute stroke are recruited into a clinical trial. All patients are treated within two hours of arrival at the emergency room with t-plasminogen activator (tPA). In addition, patients are infused with a new therapy to reduce mortality. Prior information suggests that the new therapy will profoundly amplify the therapeutic effect of tPA in improving survival.

It is anticipated that in the population of stroke patients, the three-year mortality rate is 25%. The investigators are interested in producing a risk reduction of 20%, equivalent to a 80% relative risk of mortality associated with therapy. Assuming 90% power, and a two-sided type I error rate of 0.05, the investigators require 2922 total patients, 1461 in each of the two groups. Based on this information, the total number of deaths in the study is anticipated to be $(1461)(0.25)+(1461)(0.25)(0.80)=658$.

The investigators are interested in computing the boundary values for a monitoring procedure that is based on a conditional power computation. This would require the simple application of the procedures discussed in Chapter Six. However, in the current setting, the investigators are interested in modeling the prior information for the relative risk associated with therapy $\pi(\theta)$, where

$$\pi(\theta)=\frac{a}{\theta\sqrt{2\pi}}e^{-(a\ln\theta+b)^2/2}. \tag{8.6}$$

One advantage of using this functional form for the distribution of the relative risk is that, by choosing different values of the constants a and b, one can select from a variety of shapes for the prior information (Figure 8.1).

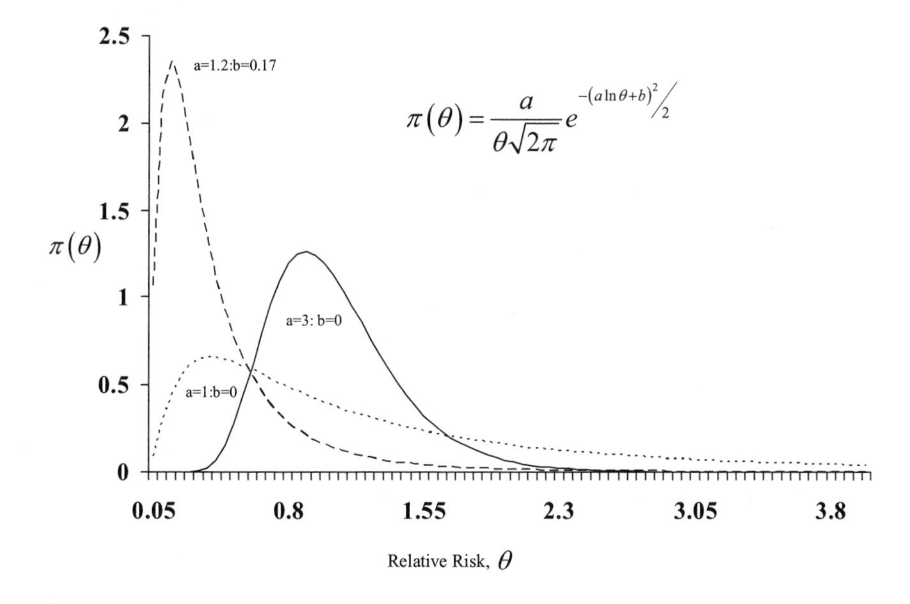

Figure 8.1. Different prior distributions for the relative risk in a clinical trial.

It is important to remember that the form of the prior distribution is chosen to represent the combination of prior beliefs about the value of the relative risk. Thus, the investigators have the freedom to calibrate $\pi(\theta)$ with their sense about the possible values of the relative risks. After much discussion among themselves they settle on the values $a = 2.9$ and $b = 0$ (Figure 8.2).

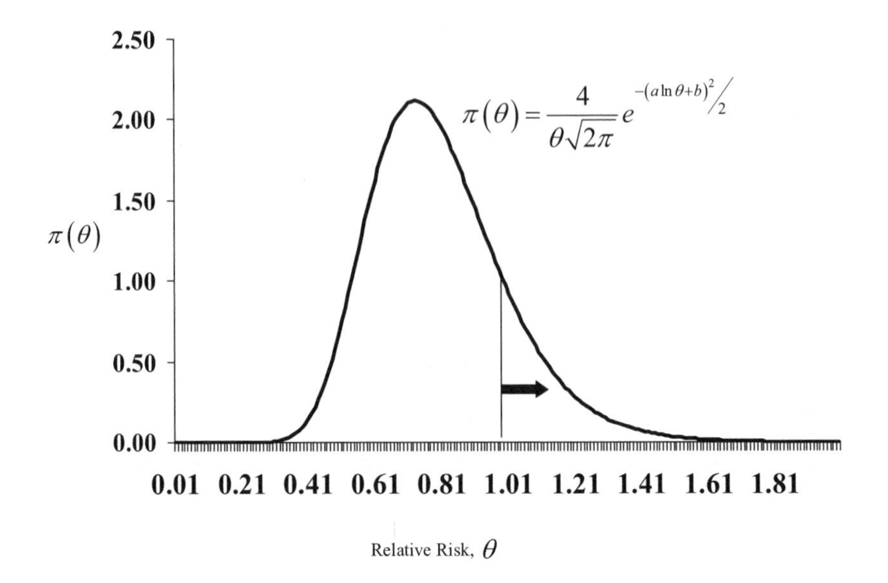

Figure 8.2. Prior distribution of relative risk of an intervention for total mortality Note the region under prior distribution that is associated with hazard (shaded arrow).

This distribution of the values of the relative risk θ in Figure 8.2 expresses the prior knowledge and beliefs of the investigators in the likely values of the relative risk. Most of the distribution of probability is amassed over values of the relative risk that are less than one, suggesting that the intervention will be effective. However, there is substantial probability associated with the likelihood that the intervention will have very different levels of effect. Note particularly the degree of probability associated with a hazard that may be produced by the intervention.

The investigators wish to compute boundary values for monitoring guidelines. Recall from Chapter Six that the boundary value computation monitoring a clinical trial was based on our ability to identify that value d such that $\boldsymbol{P}\left[TS(1)\geq Z_{1-\alpha/2} \mid TS(I_1)=d\right]=\gamma$. We see from Appendix D that the prior distribution, originally in terms of θ can be converted to be a function of the drift parameter μ. This work reveals that the prior distribution of the drift parameter μ is normal with mean $-\sqrt{E/4a^2}\,b$, and variance $\sqrt{E/4a^2}$. In this case, where $E = 658$, $b = -1$, and $a = 4.0$, the mean of the drift is -3.21, and its variance is 3.21 (Figure 8.3).

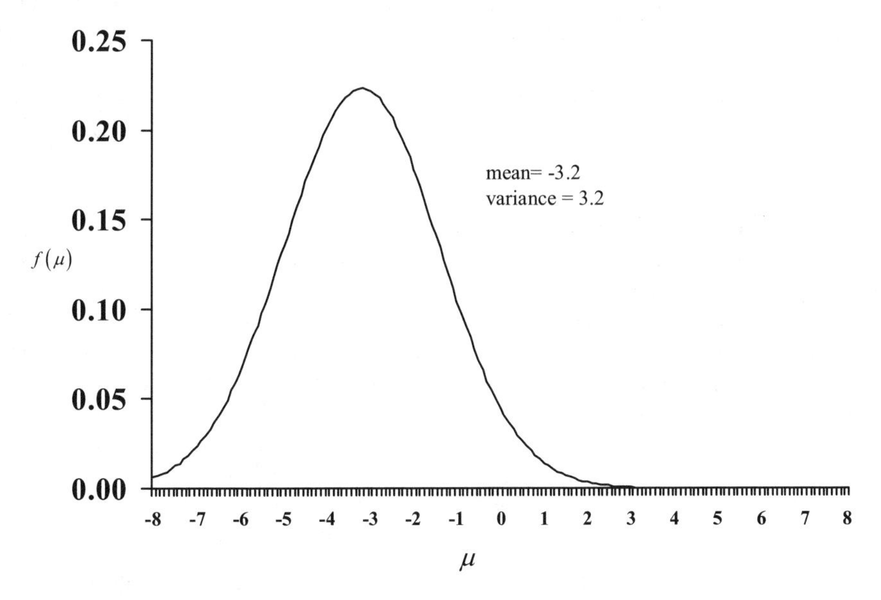

Figure 8.3. Prior probability distribution for the drift parameter in a clinical trial designed to examine stroke treatment efficacy.

We are now in a position to compute $\boldsymbol{P}\left[TS(1) \geq Z_{1-\alpha/2} \mid TS(I_1) = d\right] = \gamma$. From Appendix C, we know that

$$CP(H_0) = \gamma = \boldsymbol{P}\left[TS(1) \geq Z_{1-\alpha/2} \mid TS(I) = d\right]. \qquad (8.7)$$

Following the development in Appendix D, we find that the boundary value d is

$$d = \frac{Z_{1-\gamma}\sqrt{3.2 + (1-I)^2} \;-\; Z_{1-\alpha/2} \;+ 3.2(1-I)}{\sqrt{I}}.$$

The boundary values for this scenario can be easily computed as

$$d = \frac{1.28\sqrt{3.2 + (1-0.50)^2} \;-\; 1.96 \;+ 3.2(1-0.50)}{\sqrt{0.50}} = 2.86.$$

This is substantially lower than the value of $d = 4.05$ that is derived from a 90% conditional power computation for a two-sided type I error level of 0.05. The boundary value for the Bayesian predictive power assumption is less extreme because of the assumption that efficacy will be present for the remaining 50% of the study. This assumption of remaining efficacy means that the boundary value need not be so extreme in order for the probability to be high that the test statistic will fall in the critical region at the duration of the trial.

8.4 Mixing Prior Information

One of the important advantages in applying a Bayesian approach in statistical monitoring procedures is its ability to incorporate prior information into the computation of predictive power. However, with this flexibility comes the requirement of identifying a prior distribution that is defensible to the research and medical community. Although sometimes straightforward, this can be a complicated task.

In the previous example, we chose as a prior distribution for the therapy's efficacy,

$$\pi(\theta) = \frac{a}{\theta\sqrt{2\pi}} e^{-(a\ln\theta+b)^2/2}.$$

This distribution is remarkably malleable, as we saw from Figure 8.1, with its shape varying substantially for different values of parameters a and b. In the previous example, the investigator chose the prior distribution $a = 4$ and $b = 1$. However, the involvement of more than one investigator in the research can lead to disagreements about the content of the prior information; each may choose a different shape for the prior distribution of the level of efficacy and thus the location of θ.

Sharing all of the available information among investigators can help to resolve these differences. However, in the end, there may no single values for the parameters a and b that satisfy them. In this circumstance, it is possible to construct a mixture of prior distributions that adequately represents the collection of prior beliefs held by the investigators.

Assume the presence of two investigators 1 and 2, each of whom has his or her own values for the parameters a and b, (a_1, b_1) and (a_2, b_2): that is, they each have their own prior distribution, denoted as $\pi_1(\theta)$ and $\pi_2(\theta)$. We can define a combined or mixture prior as $\pi(\theta) = p\pi_1(\theta) + (1-p)\pi_2(\theta)$, where p is a constant between zero and one. Combining the prior distributions in this way allows each investigator's prior information to contribute to all subsequent computations involving $\pi(\theta)$. Insisting on a value of p such that $0 \le p \le 1$ assures us that the rules of probability discussed in Chapter Three remain completely intact.* In general, if there are K different prior distributions $\pi_1(\theta), \pi_2(\theta), \pi_3(\theta), \ldots, \pi_K(\theta)$, we can

* Some Bayesians dispense with the notion that the prior probabilities have to sum to one. The use of *improper priors* can in some cases be illuminating.

construct the prior distribution $\pi(\theta)=\sum_{i=1}^{K} p_i\pi_i(\theta)$ when the collection of p_i are chosen such that $\sum_{i=1}^{K} p_i = 1$. Examinations of four examples of this mixture prior $\pi(\theta)$ demonstrate that this combination process adds a new dimension of richness to the shapes of the prior distributions that are available to the investigators (Figure 8.4).

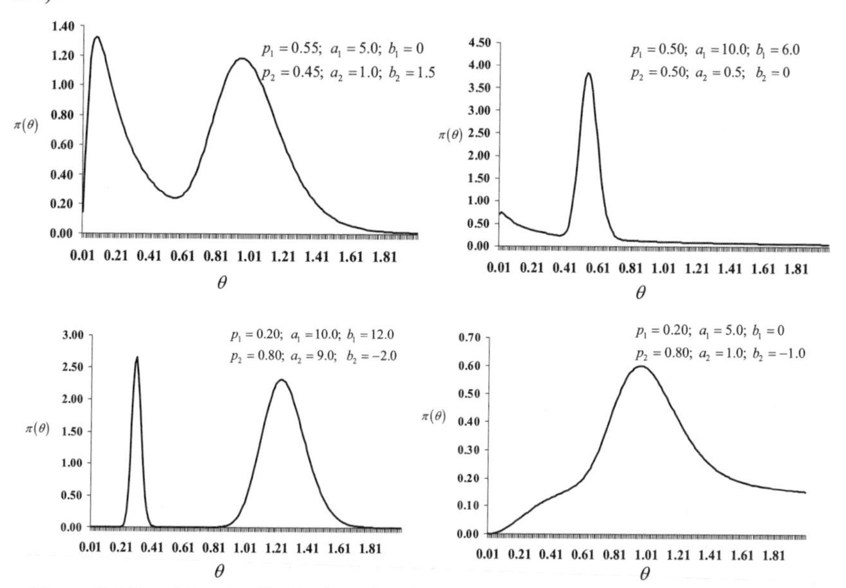

Figure 8.4. Possible prior distributions that demonstrate divergent or convergent opinions about the effect of a therapy for the acute treatment of strokes.

With this type of mixture prior distribution, the predictive power and its associated boundary values are easy to calculate. Essentially, because the prior distribution is a weighted average over each member of the collection of prior distributions, then based on our earlier work, we know that both the predictive power and the boundary values are weighed averages of their values as well.

To illustrate this process, if the investigators choose to compute the predictive power of stopping at information time I based on the value of the test statistic $TS(I) = s_1$, they would compute

$$\boldsymbol{P}\left[TS(1)\geq 1.96 | TS(I_1)=s_1\right]=\boldsymbol{P}\left[N\left(\mu(1-I_1),(1-I_1)\right)\geq 1.96-s_1\sqrt{I_1}\right].$$

Recalling that under the Bayesian paradigm, the parameter μ has a normal distribution with mean $\phi=\sqrt{E/4a^2}\,b$, and variance $\omega^2=\sqrt{E/4a^2}$, we use the result from Appendix D to simply demonstrate that the predictive power continues to be the probability of a normally distributed quantity with mean and variance related to the information time and the prior distribution. Thus, the predictive power can be written as

$$P\left[TS(1)\geq 1.96 \mid TS(I_1)=s_1\right]=1-F_Z\left[\frac{1.96-s_1\sqrt{I_1}-\sqrt{\frac{E}{4a^2}}b(1-I_1)}{\sqrt{\frac{E}{4a^2}+(1-I_1)^4}}\right]. \tag{8.8}$$

where s_1 is the value of the test statistic at information time I_1, and a and b are the parameters that the investigators chose to calibrate the prior distribution.

If the investigators instead had chosen a mixture prior distribution that produced the kinds of distributions that are depicted in Figure 8.4, then the predictive power is closely related to expression (8.8),

$$P\left[TS(1)\geq 1.96 \mid TS(I_1)=s_1\right]=\sum_{i=1}^{K}p_i\left(1-F_Z\left[\frac{1.96-s_1\sqrt{I_1}-\sqrt{\frac{E}{4a_i^2}}b_i(1-I_1)}{\sqrt{\frac{E}{4a_i^2}+(1-I_1)^4}}\right]\right),$$

which is simply a weighted sum of the predictive powers. This computation is simply the weighted sum of normal probabilities.

The idea of a mixture distribution can also be incorporated into the identification of boundary values for the interim monitoring. Recall that we find the boundary value of the test statistic d at information time I_1 by solving

$$CP(H_0)=\gamma=P\left[TS(1)\geq Z_{1-\alpha/2} \mid TS(I)=d\right]. \tag{8.9}$$

Appendix D reveals the solution for the boundary value d to be

$$d=\frac{Z_{1-\alpha/2}-Z_{1-\gamma}\sqrt{\left(\frac{E}{4a^2}+(1-I)^2\right)(1-I)}-\sqrt{\frac{E}{4a^2}}b(1-I)}{\sqrt{I}}. \tag{8.10}$$

In order to incorporate the mixture of prior probabilities as in Figure 8.4, we simply need to write expression (8.10) as

$$d = \sum_{i=1}^{K} p_i \frac{Z_{1-\alpha/2} - Z_{1-\gamma}\sqrt{\left(\frac{E}{4a_i^2} + (1-I)^2\right)(1-I)} - \sqrt{\frac{E}{4a_i^2}}b_i(1-I)}{\sqrt{I}}.$$

As an example, consider four investigators interested in executing a clinical trial to study the effect of a caffeine-ethanol in association with reduction in body temperature to reduce the severity of stroke. The investigators have widely different points of view about the effect of therapy (Figure 8.5).

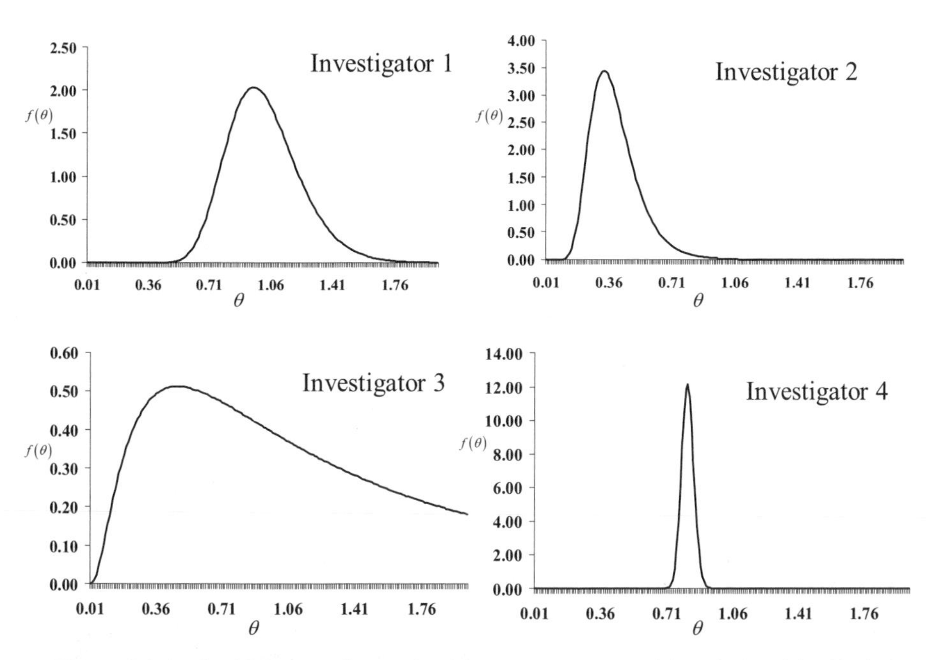

Figure 8.5. Each of four investigators has his own assessment of the relative risk afforded by the investigational therapy.

Each panel in Figure 8.5 reflects an investigator's sense of the prior information about the value of the relative risk θ. Values of θ less than one mean the therapy is beneficial, and values greater than one denote hazard. The first investigator's point of view is that there is likely to be a relatively small, beneficial effect of the drug. Investigator 2 and investigator 3 believe that the therapy is likely to be quite effective, and investigator 4's prior belief reflects high confidence in a moderate degree of efficacy.

The investigators discuss among themselves, but despite their best efforts not one of them can convince the remaining three to give up her individual prior. The researchers therefore construct a composite prior based on the following weights: investigator 1 (0.20); investigator 2 (0.40); investigator 3 (0.30), investigator (0.10). These weights are then easily used to assemble the composite prior to represent the combined sense of the investigators' prior beliefs (Figure 8.6). Note that the weights sum to one.

Examination of Figure 8.6 clearly reviews the influence of the investigators' own beliefs in the construction of the prior distribution. The composite distribution is bimodal, reflecting the strong beliefs of investigators two and three in substantial benefit afforded by the therapy as well as the sense from investigators 1 and 4 that the relative risk will be higher.

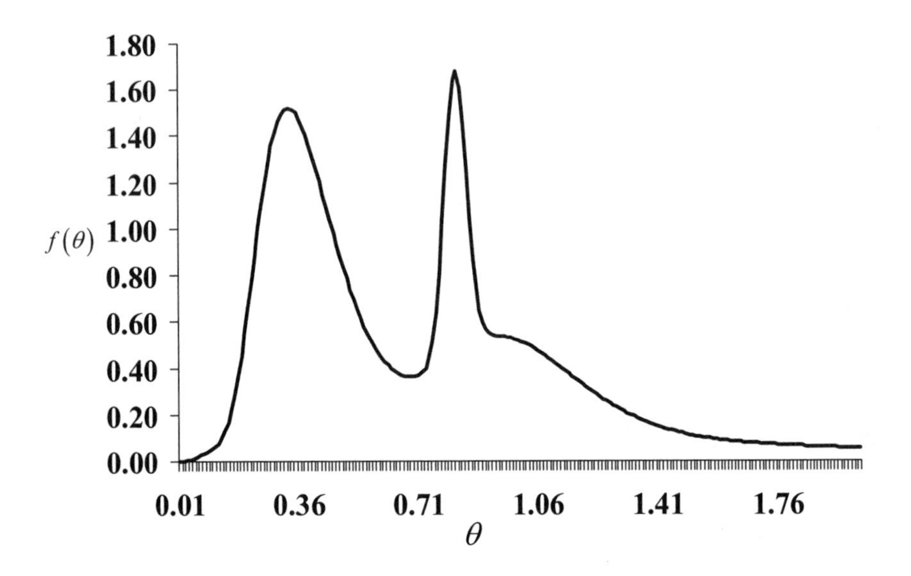

Figure 8.6. Consensus for the prior distribution for the effect of therapy.

Once the prior distribution is obtained, the investigators can use expression (8.8) to easily compute the conditional power for the value of test statistic at any information time I.

Table 8.4. Conditional Power Using a Mixture of Prior Distributions (Test Statistic = 5.1)

		Prior 1	Prior 2	Prior 3	Prior 4	**Mixture**
Information	a=	5	3	1	25	
Time	b=	0	3	–0.25	5	
0.1		0.394	0.987	0.386	0.746	**0.664**
0.2		0.606	0.992	0.446	0.953	**0.747**
0.3		0.773	0.994	0.497	0.998	**0.801**
0.4		0.883	0.994	0.541	1.000	**0.836**
0.5		0.945	0.993	0.581	1.000	**0.860**
0.6		0.975	0.991	0.617	1.000	**0.877**
0.7		0.989	0.989	0.650	1.000	**0.888**
0.8		0.995	0.985	0.681	1.000	**0.897**
0.9		0.998	0.979	0.709	1.000	**0.904**
1.0		0.999	0.970	0.735	1.000	**0.908**

8.5 Conclusions

Bayes procedures are an important new contribution to the statistical monitoring of clinical research. The incorporation of prior information provides flexibility to the investigator and a smooth, defensible, and reproducible way to incorporate the prior information into the monitoring procedure.

However, the requirement of the explicit specification of the prior distribution can be problematic if there is not much good information about the parameter to be estimated. In general, in order to be useful, prior distributions must reflect the state of information about the level of efficacy although at the same time avoiding unnecessary mathematical complexity. Like other design parameters, this should be completely pre-specified before the trial commences.

Each of the Bayes and frequentist perspectives' are useful, and one should not always be considered preferable to the other. If there is good prior information, the Bayes approach is a natural alternative. However, if there is little or no prior information, the researcher might be best served by staying with the classical perspective. In any event, scientists should make their decision prospectively.

Problems

1. Can you show that the conditional power computation in Section 8.2 is the same as the Bayesian predictive power when the prior is $\pi(\theta) = \mathbf{1}_{\theta=0}$?

2. Compute the predictive power for the monitoring situation described in Section 8.2 under the following scenarios.

Appendix A
Boundary Value for Normal Mean

Our goal here is to compute a minimum boundary value for the observed mean of a collection of normally distributed random variables based on a previously identified mean value. Specifically, let m represent the number of observations on which the interim assessment is made, such that $0 < m < n$. We know the probability distribution of $\overline{X}_n$. We need to find the probability distribution of $\overline{X}_n$ given $\overline{X}_m$. It can be demonstrated that, if $x_1, x_2, x_3, \ldots, x_p$ are independent and each follows a normal distribution with mean μ and variance σ^2, and $\overline{X}_m$ is known, $\overline{X}_n$ still follows a normal distribution but with mean μ_c, and variance v_c where*

$$\mu_c = \mu + \frac{m}{n}\left(\overline{X}_m - \mu\right)$$

$$v_c = \frac{n-m}{n^2}\sigma^2,$$

where $0 < m \leq n \leq p$. Thus, the new mean for $\overline{X}_n$ that is now conditioned on the value of $\overline{X}_m$ depends on both m and $\overline{X}_m$. The variance of $\overline{X}_n$ that is conditioned on the value of $\overline{X}_m$ depends on m.

If we let b be the upper bound of $\overline{X}_m$, we need to compute b such that

$$\boldsymbol{P}\left[\overline{X}_n \geq 67 \mid \overline{X}_m = b\right] = 0.95.$$

* This demonstration follows from a simple transformation of variables.

Under the condition $\overline{X}_m = b$, recall that the mean for $\overline{X}_n$ is $\mu + \left(m/n\right)(b-\mu)$ and its variance is $\left((n-m)/n^2\right)\sigma^2$. We may therefore, subtract this mean and divide by the standard deviation to write the preceding expression as

$$\boldsymbol{P}\left[\frac{\overline{X}_n - \left(\mu + \frac{m}{n}(b-\mu)\right)}{\sqrt{\frac{n-m}{n^2}\sigma^2}} \geq \frac{67 - \left(\mu + \frac{m}{n}(b-\mu)\right)}{\sqrt{\frac{n-m}{n^2}\sigma^2}}\right] = 0.95$$

$$\boldsymbol{P}\left[N(0,1) \geq \frac{67 - \left(\mu + \frac{m}{n}(b-\mu)\right)}{\sqrt{\frac{n-m}{n^2}\sigma^2}}\right] = 0.95.$$

Because we also know that $\boldsymbol{P}\left[N(0,1) > -1.645\right] = 0.95$, we can write

$$\frac{67 - \left(\mu + \frac{m}{n}(b-\mu)\right)}{\sqrt{\frac{n-m}{n^2}\sigma^2}} = -1.645.$$

We now solve for b

$$\frac{67 - \left(\mu + \frac{m}{n}(b-\mu)\right)}{\sqrt{\frac{n-m}{n^2}\sigma^2}} = -1.645$$

$$67 - \left(\mu + \frac{m}{n}(b-\mu)\right) = -1.645\sqrt{\frac{n-m}{n^2}\sigma^2}$$

$$\left(\mu + \frac{m}{n}(b-\mu)\right) = 67 + 1.645\sqrt{\frac{n-m}{n^2}\sigma^2}.$$

Continuing,

$$\frac{m}{n}(b-\mu)=67+1.645\sqrt{\frac{n-m}{n^2}\sigma^2}-\mu$$

$$b=\frac{n}{m}\left[67+1.645\sqrt{\frac{n-m}{n^2}\sigma^2}-\mu\right]+\mu.$$

If $\overline{X}_m \geq b$ there is a 95% probability that $\overline{X}_n \geq 67$.

Appendix B
Conditional Brownian Motion

One of the most interesting and useful features of Brownian motion is the unique nature of the dependence. This dependence can easily be characterized, and this characterization is based on the parameters of the distribution. The section will use some of the features of the multivariate normal distribution in order to identify the probability distribution of (1) Brownian motion conditioned on the past, and (2) Brownian motion conditioned on both the future and the past (Brownian bridge).

B.1 Nomenclature and Background Mathematics

The goal of this appendix is to identify the probability density function of Brownian motion conditioned on the past, and Brownian motion conditioned on both the past and the future. This work is straightforward, and for the neophyte to Brownian motion, even illuminating. However, it will require a preamble of mathematics, which is provided before derivation of the main results.

B.1.1 Notation

The results will be in terms of standard Brownian motion $B(t)$. If $x = B(t)$, then the probability density function of x is

$$f_X(x) = \frac{1}{\sqrt{2\pi t}} e^{-\frac{x^2}{2t}}. \tag{B.1}$$

As was pointed out in Chapter Three, it is the area under this curve that produces the probabilities for events involving possible interval values of x. In addition, it will be helpful to define times t_1, t_2, and t_3 such that $0 < t_1 \leq t_2 \leq t_3$. Define $x_1 = B(t_1)$, $x_2 = B(t_2)$, and $x_3 = B(t_3)$.

B.1.2 Multivariate Normal Distributions

Two random variables W and V follow a joint normal distribution if the function that governs their probability computations (i.e., the probability density function) is

$$f_{W,V}(w,v) = \frac{1}{2\pi \det(\Sigma)^{1/2}} e^{-\frac{1}{2}(w-\mu_w \quad v-\mu_v)\Sigma^{-1}\begin{pmatrix} w-\mu_w \\ v-\mu_v \end{pmatrix}} \tag{B.2}$$

where the mean of W is μ_W, the mean of V is μ_V. The quantity Σ is known as the 2 x 2 variance-covariance matrix of the random variables W and V when they are assembled into a 2 by 1 vector $\begin{bmatrix} W \\ V \end{bmatrix}$. In the circumstance of two random variables,

$$\Sigma = \begin{bmatrix} Var(W) & Cov(W,V) \\ Cov(W,V) & Var(V) \end{bmatrix}.$$

In the cases that will be evaluated in this appendix $\mu_W = \mu_V = 0$. In this case equation (B.2) may be written as

$$f_{W,V}(w,v) = \frac{1}{2\pi \det(\Sigma)^{1/2}} e^{-\frac{1}{2}(w \quad v)\Sigma^{-1}\begin{pmatrix} w \\ v \end{pmatrix}} \tag{B.3}$$

In the case of three jointly distributed normal random variables, W, V, and Y, each with means of zero, we may generalize equation (B.3) to write

$$f_{W,V,Y}(w,v,y) = \frac{1}{2\pi \det(\Sigma)^{1/2}} e^{-\frac{1}{2}(w \quad v \quad y)\Sigma^{-1}\begin{pmatrix} w \\ v \\ y \end{pmatrix}} \tag{B.4}$$

where

$$\Sigma = \begin{bmatrix} Var(W) & Cov(W,V) & Cov(W,Y) \\ Cov(W,V) & Var(V) & Cov(V,Y) \\ Cov(W,Y) & Cov(V,Y) & Var(Y) \end{bmatrix} \tag{B.5}$$

B.2 Covariance and Brownian Motion Elements

If we are to apply formulas (B.3), and (B.4) to the joint distribution of Brownian motion elements, we will need to identify the elements of the variance-covariance matrix Σ. This requires us to find the covariance between two Brownian elements.

As defined in section A.1. let t_1, t_2 be times such that $0 < t_1 \le t_2$, and define $x_1 = B(t_1)$, $x_2 = B(t_2)$. We must identify the covariance of x_1 and x_2. This covariance is defined as

$$Cov(X_1, X_2) = E[X_1X_2] - E[X_1]E[X_2] = E[X_1X_2] \tag{B.6}$$

where E[W] is the expected value or mean of the random variable W. The right-most equality stems from the fact that in the circumstance of standard Brownian motion, $E[X_1] = E[X_2] = 0$.

To compute $E[X_1X_2]$, we use a double expectation argument, i.e., $E[g(X_1, X_2)] = E_{X_1}[E_{X_2}[g(X_1, X_2) \mid X_1 = x_1]]$. Applying this result to $g(X_1, X_2) = X_1X_2$, and computing the inner expectation first, we find

$$E_{X_2}[X_1X_2 \mid X_1 = x_1] = E_{X_2}[x_1X_2 \mid X_1 = x_1] = (x_1)E_{X_2}[X_2 \mid X_1 = x_1] = (x_1)(x_1) = x_1^2.$$

Continuing,

$$Cov(X_1, X_2) = E[X_1X_2] = E_{X_1}[E_{X_2}[X_1X_2 \mid X_1 = x_1]] = E_{X_1}[x_1^2] = t_1.$$

This last result follows from $t_1 = Var(X_1) = E[x_1^2] - (E[x_1])^2 = E[x_1^2]$. Thus, the covariance of two Brownian elements $x_1 = B(t_1)$ and $x_2 = B(t_2)$ is the minimum of the times t_1 and t_2.

The correlation between these elements follows.

$$Corr(X_1, X_2) = \frac{Cov(X_1, X_2)}{\sigma_{X_1}, \sigma_{X2}} = \frac{t_1}{\sqrt{t_1t_2}} = \sqrt{\frac{t_1}{t_2}}.$$

where $0 \le t_1 \le t_2$. Note that when $t_1 = 0$, the correlation between $B(0)$ and $B(t_2)$ is zero. This correlation increases as t_1 approaches t_2, reaching its maximum value of one when $t_1 = t_2$.

B.3 Brownian Motion Conditioned on the Past

With this as background, we can now find the distribution of a Brownian motion element that is conditioned on the past. Let $0 < t_1 \le t_2$ and $x_1 = B(t_1)$, $x_2 = B(t_2)$. We are interested in the probability distribution of x_2 given x_1. Relying on the discussions in Chapter Three, we write

$$f(x_2 \mid x_1) = \frac{f_{X_1,X_2}(x_1, x_2)}{f_{X_1}(x_1)}. \tag{B.7}$$

We know that $f_{X_1}(x_1) = \frac{1}{\sqrt{2\pi t_1}} e^{-\frac{x_1^2}{2t_1}}$. Using equation (B.3), we write that

$$f_{X_1,X_2}(x_1,x_2) = \frac{1}{2\pi \det(\Sigma)^{1/2}} e^{-\frac{1}{2}(x_1 \quad x_2)\Sigma^{-1}\begin{pmatrix} x_1 \\ x_2 \end{pmatrix}} \tag{B.8}$$

From our examination of the covariance between two Brownian motion elements, we know from the previous section that

$$\Sigma = \begin{bmatrix} Var(X_1) & Cov(X_1,X_2) \\ Cov(X_1,X_2) & Var(X_2) \end{bmatrix} = \begin{bmatrix} t_1 & t_1 \\ t_1 & t_2 \end{bmatrix}. \tag{B.9}$$

The $\det(\Sigma) = t_1 t_2 - t_1^2 = t_1(t_2 - t_1)$, and we can write

$$\Sigma^{-1} = \begin{bmatrix} t_1 & t_1 \\ t_1 & t_2 \end{bmatrix}^{-1} = \begin{bmatrix} t_2 & -t_1 \\ -t_1 & t_1 \end{bmatrix} \frac{1}{t_1(t_2 - t_1)} \tag{B.10}$$

Applying these results to the exponent of equation (B.8), we have

$$\begin{aligned} -\frac{1}{2}(x_1 \quad x_2)\Sigma^{-1}\begin{pmatrix} x_1 \\ x_2 \end{pmatrix} &= -\frac{1}{2}(x_1 \quad x_2)\begin{bmatrix} t_2 & -t_1 \\ -t_1 & t_1 \end{bmatrix}\frac{1}{t_1(t_2 - t_1)}\begin{pmatrix} x_1 \\ x_2 \end{pmatrix} \\ &= -\frac{1}{2}\frac{t_1 x_2^2 - 2t_1 x_1 x_2 + t_2 x_1^2}{t_1(t_2 - t_1)} \end{aligned} \tag{B.11}$$

We can then write the conditional density of x_2 given x_1 as

$$\begin{aligned} f(x_2 \mid x_1) = \frac{f_{X_1,X_2}(x_1,x_2)}{f_{X_1}(x_1)} &= \frac{\frac{1}{2\pi\sqrt{t_1(t_2 - t_1)}} e^{-\frac{1}{2}\frac{t_1 x_2^2 - 2t_1 x_1 x_2 + t_2 x_1^2}{t_1(t_2 - t_1)}}}{\frac{1}{\sqrt{2\pi t_1}} e^{-\frac{x_1^2}{2t_1}}} \\ &= \frac{1}{\sqrt{2\pi(t_2 - t_1)}} e^{-\frac{1}{2}\left[\frac{t_1 x_2^2 - 2t_1 x_1 x_2 + t_2 x_1^2}{t_1(t_2 - t_1)} + \frac{x_1^2}{t_1}\right]}. \end{aligned} \tag{B.12}$$

Examining the exponent of the last line of expression (B.12), we find

$$
\begin{aligned}
&-\frac{1}{2}\left[\frac{t_1x_2^2-2t_1x_1x_2+t_2x_1^2}{t_1(t_2-t_1)}+\frac{x_1^2}{t_1}\right] \\
&= -\frac{1}{2(t_2-t_1)}\left[x_2^2-2x_1x_2+\frac{t_2}{t_1}x_1^2\right]-\frac{x_1^2}{2t_1} \\
&= -\frac{1}{2(t_2-t_1)}\left[x_2^2-2x_1x_2\right]-\frac{t_2x_1^2}{2t_1(t_2-t_1)}-\frac{x_1^2}{2t_1}.
\end{aligned}
\tag{B.13}
$$

Completing the square for the term $x_2^2-2x_1x_2$ reveals

$$
\begin{aligned}
&-\frac{1}{2(t_2-t_1)}\left[x_2^2-2x_1x_2+x_1^2-x_1^2\right]-\frac{t_2x_1^2}{2t_1(t_2-t_1)}-\frac{x_1^2}{2t_1} \\
&= -\frac{1}{2(t_2-t_1)}\left[x_2^2-2x_1x_2+x_1^2\right]+\frac{x_1^2}{2(t_2-t_1)}-\frac{t_2x_1^2}{2t_1(t_2-t_1)}-\frac{x_1^2}{2t_1} \\
&= -\frac{1}{2(t_2-t_1)}(x_2-x_1)^2.
\end{aligned}
$$

With this simplification of the exponent, we may now write the conditional density of x_2 given x_1.

$$
f(x_2 \mid x_1)=\frac{1}{\sqrt{2\pi(t_2-t_1)}}e^{-\frac{1}{2(t_2-t_1)}(x_2-x_1)^2}
\tag{B.14}
$$

This is the density of a normal random variable with mean x_1 and variance $t_2 - t_1$.

B.4 Brownian Motion Conditioned on the Past and Future

We are now in a position to find the distribution of a Brownian element that is conditioned both on the past and the future. In this circumstance, we are given and $x_1 = B(t_1)$, $x_2 = B(t_2)$, and $x_3 = B(t_3)$ where $0 < t_1 < t_2 < t_3$. Following the previous development of section A.3, we

$$
f(x_2 \mid x_1,x_3)=\frac{f_{X_1,X_2,X_3}(x_1,x_2,x_3)}{f_{X_1,X_3}(x_1,x_3)}.
\tag{B.15}
$$

Using equation (B.3), we are able to write the denominator of this conditional probability distribution as that $f_{X_1,X_3}(x_1,x_3)=\frac{1}{2\pi\sqrt{t_1(t_3-t_1)}}e^{-\frac{1}{2}\frac{t_1x_3^2-2t_1x_1x_3+t_2x_1^2}{t_1(t_3-t_1)}}$. We proceed with developing the numerator of expression (B.15)

$$f_{X_1,X_2,X_3}(x_1,x_2,x_3)=\frac{1}{(2\pi)^{\frac{3}{2}}\det(\Sigma)^{1/2}}e^{-\frac{1}{2}(x_1 \quad x_2 \quad x_3)\Sigma^{-1}\begin{pmatrix}x_1\\x_2\\x_3\end{pmatrix}}. \tag{B.16}$$

From our examination of the covariance between two Brownian motion elements, we know from the previous section that

$$\Sigma=\begin{bmatrix}Var(X_1) & Cov(X_1,X_2) & Cov(X_1,X_3)\\ Cov(X_1,X_3) & Var(X_2) & Cov(X_2,X_3)\\ Cov(X_1,X_3) & Cov(X_2,X_3) & Var(X_3)\end{bmatrix}=\begin{bmatrix}t_1 & t_1 & t_1\\ t_1 & t_2 & t_2\\ t_1 & t_2 & t_3\end{bmatrix}. \tag{B.17}$$

The $\det(\Sigma)=t_1(t_3-t_2)(t_2-t_1)$. Using elementary row and column operations, we can write

$$\Sigma^{-1}=\begin{bmatrix}t_1 & t_1 & t_1\\ t_1 & t_2 & t_2\\ t_1 & t_2 & t_3\end{bmatrix}^{-1}=\begin{bmatrix}\frac{1}{t_1}+\frac{1}{t_2-t_1} & \frac{-1}{t_2-t_1} & 0\\ \frac{-1}{t_2-t_1} & \frac{1}{t_2-t_1}+\frac{1}{t_3-t_2} & \frac{-1}{t_3-t_2}\\ 0 & \frac{-1}{t_3-t_2} & \frac{1}{t_3-t_2}\end{bmatrix} \tag{B.18}$$

Applying these results to the exponent of equation (B.16), we have

$$\begin{aligned}&-\frac{1}{2}(x_1 \quad x_2 \quad x_3)\Sigma^{-1}\begin{pmatrix}x_1\\x_2\\x_3\end{pmatrix}\\ &=-\frac{1}{2}(x_1 \quad x_2 \quad x_3)\begin{bmatrix}\frac{1}{t_1}+\frac{1}{t_2-t_1} & \frac{-1}{t_2-t_1} & 0\\ \frac{-1}{t_2-t_1} & \frac{1}{t_2-t_1}+\frac{1}{t_3-t_2} & \frac{-1}{t_3-t_2}\\ 0 & \frac{-1}{t_3-t_2} & \frac{1}{t_3-t_2}\end{bmatrix}\begin{pmatrix}x_1\\x_2\\x_3\end{pmatrix}\\ &=-\frac{1}{2}\left[\left(\frac{1}{t_2-t_1}+\frac{1}{t_3-t_2}\right)x_2^2-2\left(\frac{x_1}{t_2-t_1}+\frac{x_3}{t_3-t_2}\right)x_2+\left(\frac{1}{t_1}+\frac{1}{t_2-t_1}\right)x_1^2++\left(\frac{1}{t_3-t_2}\right)x_3^2\right]\end{aligned} \tag{B.19}$$

We can then write the conditional density of x_2 given x_1 and x_3 as

$$f\left(x_2 \mid x_1, x_3\right) = \frac{f_{X_1,X_2,X_3}\left(x_1,x_2,x_3\right)}{f_{X_1,X_3}\left(x_1,x_3\right)}$$

$$= \frac{\dfrac{1}{(2\pi)^{\frac{3}{2}}\sqrt{t_1\left(t_3-t_2\right)\left(t_2-t_1\right)}} e^{-\frac{1}{2}\left[\left(\frac{1}{t_2-t_1}+\frac{1}{t_3-t_2}\right)x_2^2-2\left(\frac{x_1}{t_2-t_1}+\frac{x_3}{t_3-t_2}\right)x_2+\left(\frac{1}{t_1}+\frac{1}{t_2-t_1}\right)x_1^2+\left(\frac{1}{t_3-t_2}\right)x_3^2\right]}}{\dfrac{1}{2\pi\sqrt{t_1\left(t_3-t_1\right)}} e^{-\frac{1}{2}\frac{t_1x_3^2-2t_1x_1x_3+t_2x_1^2}{t_1\left(t_3-t_1\right)}}}$$

. (B.20)

This last expression requires simplification. Begin by writing

$$\frac{\dfrac{1}{(2\pi)^{\frac{3}{2}}\sqrt{t_1\left(t_3-t_2\right)\left(t_2-t_1\right)}}}{\dfrac{1}{2\pi\sqrt{t_1\left(t_3-t_1\right)}}} = \frac{1}{\sqrt{2\pi}}\frac{\sqrt{t_1\left(t_3-t_1\right)}}{\sqrt{t_1\left(t_3-t_2\right)\left(t_2-t_1\right)}} = \frac{1}{\sqrt{2\pi}\sqrt{\dfrac{\left(t_3-t_2\right)\left(t_2-t_1\right)}{\left(t_3-t_1\right)}}}$$

This simplification reveals that the variance of the conditional distribution of x_2 given x_1 and x_3 will be $\frac{\left(t_3-t_2\right)\left(t_2-t_1\right)}{\left(t_3-t_1\right)}$. We can use this observation to guide our simplification of the exponent. Begin by writing the exponent as

$$\left(\frac{1}{t_2-t_1}+\frac{1}{t_3-t_2}\right)x_2^2-2\left(\frac{x_1}{t_2-t_1}+\frac{x_3}{t_3-t_2}\right)x_2+\left(\frac{1}{t_1}+\frac{1}{t_2-t_1}\right)x_1^2+\left(\frac{1}{t_3-t_2}\right)x_3^2$$
$$+\frac{t_1x_3^2-2t_1x_1x_3+t_2x_1^2}{t_1\left(t_3-t_1\right)}$$

Note that only the first two terms contain the variable x_2. Completing the square with respect to x_2 and simplifying reveals that the exponent can be written as

$$-\frac{1}{2}\left(\frac{\left(t_3-t_1\right)}{\left(t_3-t_2\right)\left(t_2-t_1\right)}\right)\left[x_2-\frac{\left(t_3-t_2\right)x_1+\left(t_2-t_1\right)x_3}{t_3-t_1}\right]^2.$$

Thus, the conditional probability can be written as

$$f(x_2 \mid x_1, x_3) = \frac{1}{\sqrt{2\pi}\sqrt{\frac{(t_3 - t_2)(t_2 - t_1)}{(t_3 - t_1)}}} e^{-\frac{1}{2}\left(\frac{(t_3-t_1)}{(t_3-t_2)(t_2-t_1)}\right)\left[x_2 - \frac{(t_3-t_2)x_1+(t_2-t_1)x_3}{t_3-t_1}\right]^2}.$$

This is a normal distribution with mean $\mu_c = \frac{(t_3 - t_2)x_1 + (t_2 - t_1)x_3}{t_3 - t_1}$ and variance $v_c = \frac{(t_3 - t_2)(t_2 - t_1)}{(t_3 - t_1)}$.

Appendix C
Boundary Values and Conditional Power

Our goal here is to compute a minimum boundary value for efficacy/harm and the boundary value for stopping a clinical research effort because of the all but inevitable absence of clinical effect (futility). These computations will be carried out under conditional power assumptions.

C.1 Monitoring for Efficacy or Harm

In this setting, the investigators wish to compute the value b such that, if the test statistic at information time I is equal to b, then the probability that the test statistic falls in the critical region at the conclusion of the study is at least some value γ when γ is large (i.e., $\gamma = 0.95$). The probability statement for this occurrence is

$$CP(H_0) = \gamma = \boldsymbol{P}\left[TS(1) \geq Z_{1-\alpha/2} \mid TS(I) = b\right] \tag{C.1}$$

To solve for b, we first convert expression (C.1) to a statement involving a Brownian motion event.

$$\gamma = \boldsymbol{P}\left[B(1) \geq Z_{1-\alpha/2} \mid B(I) = \sqrt{I}b\right]$$

Recognizing this as Brownian motion conditioned on the past can write.

$$\gamma = \boldsymbol{P}\left[B(1-I) \geq Z_{1-\alpha/2} - \sqrt{I}b\right]$$

Because the investigator is working under the null hypothesis, this expression may be written as

$$\gamma = \boldsymbol{P}\left[N(0,1-I) \geq Z_{1-\alpha/2} - \sqrt{I}b\right]$$
$$= \boldsymbol{P}\left[N(0,1) \geq \frac{Z_{1-\alpha/2} - \sqrt{I}b}{\sqrt{1-I}}\right].$$

However, because $\gamma = \boldsymbol{P}\left[N(0,1) \geq Z_{1-\gamma}\right]$, we may now write

$$Z_{1-\gamma} = \frac{Z_{1-\alpha/2} - \sqrt{I_1}b}{\sqrt{1-I_1}} \tag{C.2}$$

Solving expression (C.2) for b reveals

$$b = \frac{Z_{1-\alpha/2} - Z_{1-\gamma}\sqrt{1-I_1}}{\sqrt{I_1}}. \tag{C.3}$$

The probability statement for this occurrence for harm is

$$CP(H_0) = \gamma = \boldsymbol{P}\left[TS(1) \leq Z_{\alpha/2} \mid TS(I) = b\right] \tag{C.4}$$

To solve for b, we first convert expression (C.1) to a statement involving a Brownian motion event.

$$\gamma = \boldsymbol{P}\left[B(1) \leq Z_{\alpha/2} \mid B(I) = \sqrt{I}b\right]$$

Recognizing this as Brownian motion conditioned on the past can write.

$$\gamma = \boldsymbol{P}\left[B(1-I) \leq Z_{\alpha/2} - \sqrt{I}b\right]$$

Because the investigator is working under the null hypothesis, this expression may be written as

$$\gamma = \boldsymbol{P}\left[N(0,1-I) \leq Z_{\alpha/2} - \sqrt{I}b\right]$$
$$= \boldsymbol{P}\left[N(0,1) \leq \frac{Z_{\alpha/2} - \sqrt{I}b}{\sqrt{1-I}}\right].$$

However, because $\gamma = \boldsymbol{P}\left[N(0,1) \leq Z_{\gamma}\right]$, we may now write

$$Z_\gamma = \frac{Z_{\alpha/2} - \sqrt{I}b}{\sqrt{1-I}} \tag{C.5}$$

Solving expression (C.2) for b reveals

$$b = \frac{Z_{\alpha/2} - Z_\gamma \sqrt{1-I}}{\sqrt{I}}. \tag{C.6}$$

C.2 Calculation for Futility

A similar style of calculation may be carried out to compute the value of a boundary value such that, if achieved, would reflect the low likelihood of a positive research result even under the most optimistic case. In this circumstance, the investigators are concerned about the relatively low probability that the test statistic will fall in the critical region at the end of the study, given the test statistic at information time I_1 is equal to a value f. In the conditional power environment, this may be written as

$$CP(H_a) = \upsilon = \boldsymbol{P}\left[TS(1) \geq Z_{1-\alpha/2} \mid TS(I_1) = f\right]. \tag{C.7}$$

where v is a small value (i.e., $v = 0.10$).

We proceed as we did before, converting the event in expression (C.7) to an equivalent event involving Brownian motion, then invoking the property of Brownian motion conditioned on the past.

$$\begin{aligned} \upsilon &= \boldsymbol{P}\left[TS(1) \geq Z_{1-\alpha/2} \mid \sqrt{I_1}\, TS(I_1) = f\sqrt{I_1}\right] \\ &= \boldsymbol{P}\left[B(1) \geq Z_{1-\alpha/2} \mid B(I_1) = f\sqrt{I_1}\right] \\ &= \boldsymbol{P}\left[B(1-I_1) \geq Z_{1-\alpha/2} - f\sqrt{I_1}\right]. \end{aligned} \tag{C.8}$$

At this point, we note that under the alternative hypothesis, the Brownian process has a nonzero drift parameter μ. We may now rewrite the last line of expression (C.8) as

$$\upsilon = \boldsymbol{P}\left[N\left(\mu(1-I_1), 1-I_1\right) \geq Z_{1-\alpha/2} - f\sqrt{I_1}\right],$$

standardizing to

$$\upsilon = \boldsymbol{P}\left[N(0,1) \geq \frac{Z_{1-\alpha/2} - f\sqrt{I_1} - \mu(1-I_1)}{\sqrt{1-I_1}}\right]. \tag{C.9}$$

The probability of the event in expressed in (C.9) is low. We may therefore, write

$\upsilon = \boldsymbol{P}\left[N(0,1) \geq Z_{1-\upsilon}\right]$. This statement, in combination with (C.9) reveals

$$Z_{1-\upsilon} = \frac{Z_{1-\alpha/2} - f\sqrt{I_1} - \mu(1-I_1)}{\sqrt{1-I_1}},$$

which, when solved for f, reveals

$$f = \frac{Z_{1-\alpha/2} - Z_{1-\upsilon}\sqrt{1-I_1} - \mu(1-I_1)}{\sqrt{I_1}}. \tag{C.10}$$

A similar computation can be carried out for futility examining the issue of safety. In this circumstance, the investigators are concerned about the relatively low probability that the test statistic monitoring safety will fall in the critical region at the end of the study, given the test statistic at information time I_1 is equal to a value f_s. where f_s denotes the futility value for safety. In the conditional power environment, this may be written as

$$CP(H_a) = \upsilon_f = \boldsymbol{P}\left[TS(1) \leq Z_{\alpha/2} \mid TS(I_1) = f_s\right]. \tag{C.11}$$

where v_f denotes conditional power for safety, and, as for the futility computation for efficacy is a small value.

We proceed as we did before, converting the event in expression (C.7) to an equivalent event involving Brownian motion, then invoking the property of Brownian motion conditioned on the past.

$$\begin{aligned}
\upsilon_f &= \boldsymbol{P}\left[TS(1) \leq Z_{\alpha/2} \mid \sqrt{I_1}\, TS(I_1) = f_s\sqrt{I_1}\right] \\
&= \boldsymbol{P}\left[B(1) \leq Z_{\alpha/2} \mid B(I_1) = f_s\sqrt{I_1}\right] \\
&= \boldsymbol{P}\left[B(1-I_1) \leq Z_{\alpha/2} - f_s\sqrt{I_1}\right].
\end{aligned} \tag{C.12}$$

At this point, we note that under the alternative hypothesis, the Brownian process has a nonzero drift parameter μ. We may now rewrite the last line of expression (C.8) as

$$\upsilon = \boldsymbol{P}\left[N\left(\mu(1-I_1), 1-I_1\right) \leq Z_{\alpha/2} - f\sqrt{I_1}\right],$$

standardizing to

$$\upsilon_s = \boldsymbol{P}\left[N(0,1) \leq \frac{Z_{\alpha/2} - f_s\sqrt{I_1} - \mu(1-I_1)}{\sqrt{1-I_1}} \right]. \tag{C.13}$$

The probability of the event in expressed in (C.9) is low. We may therefore, write $\upsilon_s = \boldsymbol{P}\left[N(0,1) \leq Z_{\upsilon_s} \right]$. This statement, in combination with (C.9) reveals

$$Z_{\upsilon_s} = \frac{Z_{\alpha/2} - f_s\sqrt{I_1} - \mu(1-I_1)}{\sqrt{1-I_1}},$$

which, when solved for f_s, reveals

$$f = \frac{Z_{\alpha/2} - Z_{\upsilon_s}\sqrt{1-I_1} - \mu(1-I_1)}{\sqrt{I_1}}. \tag{C.14}$$

Appendix D
Supporting Bayesian Computations

In Chapter Eight, we focused on the computation of a boundary value for a test statistic in a clinical trial with total mortality as a primary endpoint, and a prior distribution $\pi(\theta)$

$$\pi(\theta)=\frac{a}{\theta\sqrt{2\pi}}e^{-(a\ln\theta+b)^2/2}.$$

Investigators choose the values of the constants a and b consistent with the current medical knowledge. Our goal is to incorporate this prior distribution into a conditional power computation that would produce the boundary values that would guide interim monitoring of the trial.

Recall from Appendix C that, if the investigators wish to compute boundary value d such that, if the test statistic at information time I is equal to d, then the probability that the test statistic falls in the critical region at the conclusion of the study is at least some value γ when γ is large (i.e., $\gamma = 0.95$). The probability statement for this occurrence is

$$CP(H_0)=\gamma=\boldsymbol{P}\left[TS(1)\geq Z_{1-\alpha/2}\,|\,TS(I)=d\right]. \tag{D.1}$$

To solve for d, we first convert expression (D.1) to a statement involving a Brownian motion event.

$$\gamma=\boldsymbol{P}\left[B(1)\geq Z_{1-\alpha/2}\,|\,B(I)=\sqrt{I}\,d\right].$$

Recognizing this as Brownian motion conditioned on the past can write.

$$\gamma=\boldsymbol{P}\left[B(1-I)\geq Z_{1-\alpha/2}-\sqrt{I}\,d\right]. \tag{D.2}$$

The traditional conditional power approach assumes that there is no efficacy for the unexpired duration of the study. However, in the Bayesian setting, we assume that there is efficacy for the remainder of the study at a level dictated by the prior distribution $\pi(\theta)$. Let the drift parameter be μ. Then we may transform expression (D.2) to

$$\gamma = \mathbf{P}\left[N\left(\mu(1-I),1-I\right) \geq Z_{1-\alpha/2} - \sqrt{I}\, d \right]. \tag{D.3}$$

This probability can be evaluated conditional on the value of μ. However, we only have the prior distribution in terms of the relative risk θ. Our plan is to link μ to θ for which we have a prior distribution. Recall from Chapter Six that the drift parameter μ can be written in terms of the relative risk θ as

$$\mu = -\ln\theta\sqrt{\frac{E}{4}}, \tag{D.4}$$

where E is the anticipated total number of primary endpoint events in the trial. To convert prior probability distribution for θ, $\pi(\theta)$ to the probability distribution to μ, $\pi(\mu)$ using expression recall that

$$\pi(\theta) = \frac{a}{\theta\sqrt{2\pi}} e^{-(a\ln\theta+b)^2/2}. \tag{D.5}$$

Begin by writing $w = \ln(\theta)$. Then $\theta = e^w$, and $d\theta = e^w dw$. Then we may write

$$\begin{aligned}\pi(w) &= \frac{a}{e^w\sqrt{2\pi}} e^{-\left(a\ln e^w + b\right)^2/2} e^w \\ &= \frac{a}{\sqrt{2\pi}} e^{-(aw+b)^2/2}.\end{aligned}$$

Letting $\mu = -\sqrt{\frac{E}{4}}\, w$, we may transform $\pi(w)$ to $\pi(\mu)$.

$$\pi(w) = \frac{a}{\sqrt{2\pi}} e^{-\left(a\sqrt{\frac{4}{E}}\mu + b\right)^2 \Big/ 2} \sqrt{\frac{4}{E}}$$

$$= \frac{1}{\sqrt{2\pi \frac{E}{4a^2}}} e^{\frac{-1}{2\frac{E}{4a^2}}\left(\mu - \sqrt{\frac{E}{4a^2}}b\right)^2}.$$

Thus, the prior distribution of μ, $\pi(\mu)$ is normal with mean $\sqrt{E/4a^2}\,b$ and variance $\sqrt{E/4a^2}$.

This result must now be used to identify the distribution of Brownian motion with mean $\mu(1-I)$ and variance $(1-I)$ where μ itself is a random variable following a normal distribution with mean $\sqrt{E/4a^2}\,b$ and variance $\sqrt{E/4a^2}$.

We will solve the problem in general. Let X be a variable that follows a normal distribution with mean $\mu(1-I)$ and variance σ^2. Let μ itself follow a normal distribution with mean ϕ and variance ω^2. We must identify the unconditional probability for X. From $f_X(x) = \int f_X(x \mid \mu)\pi(u)$, we may write

$$f_X(x) = \int f_X(x \mid \mu)\pi(u)$$

$$= \int_{-\infty}^{\infty} \frac{1}{\sqrt{2\pi\sigma^2}} e^{-\frac{(x-\mu(1-I))^2}{2\sigma^2}} dx \frac{1}{\sqrt{2\pi\omega^2}} e^{-\frac{(\mu-\phi)^2}{2\omega^2}} d\mu \qquad \text{(D.6)}$$

$$= \int_{-\infty}^{\infty} \frac{1}{\sqrt{2\pi\sigma^2\omega^2}} e^{-\left[\frac{(x-\mu(1-I))^2}{2\sigma^2} + \frac{(\mu-\phi)^2}{2\omega^2}\right]} dx d\mu$$

Our goal will be to complete the square in the exponential, adding the terms that will allow us to carry out the integration with respect to μ. It will then remain for us to recognize the remaining terms which will be in terms of φ, σ^2, and ω^2. We proceed by defining the exponent in the last line of expression (D.6) as A and writing

$$A = -\left[\frac{(x-\mu(1-I))^2}{2\sigma^2} + \frac{(\mu-\phi)^2}{2\omega^2}\right] = -\frac{1}{2}\left[\frac{(x-\mu(1-I))^2}{\sigma^2} + \frac{(\mu-\phi)^2}{\omega^2}\right]$$

$$= -\frac{1}{2}\left[\frac{\omega^2(x-\mu(1-I))^2}{\sigma^2\omega^2} + \frac{\sigma^2(\mu-\phi)^2}{\sigma^2\omega^2}\right].$$

Continuing,

$$A = \frac{-1}{2\sigma^2\omega^2}\left[\omega^2(x-\mu(1-I))^2 + \sigma^2(\mu-\phi)^2\right]$$

$$= \frac{-1}{2\sigma^2\omega^2}\left[\omega^2(x-\mu(1-I))^2 + \sigma^2(\mu-\phi)^2\right]$$

$$= \frac{-1}{2\sigma^2\omega^2}\left[\left(\omega^2 + (\sigma(1-I))^2\right)\mu^2 - 2\left(\omega^2\phi + \sigma^2(1-I)x\right)\mu + \omega^2\phi^2 + \sigma^2x^2\right]$$

Define

$$\rho = \frac{\sigma^2\omega^2}{\left(\omega^2 + (\sigma(1-I))^2\right)}$$

Continuing, with the goal of substituting the value one for the coefficient of μ^2, we find

$$A = \frac{-1}{2\rho}\left[\mu^2 - 2\frac{\left(\omega^2\phi + \sigma^2(1-I)x\right)}{\omega^2 + (\sigma(1-I))^2}\mu + \frac{\omega^2\phi^2 + \sigma^2x^2}{\omega^2 + (\sigma(1-I))^2}\right]$$

which may be rewritten as

$$A = \frac{-1}{2\rho}\left[\mu^2 - 2\frac{\left(\omega^2\phi + \sigma^2(1-I)x\right)}{\omega^2 + (\sigma(1-I))^2}\mu\right] - \frac{1}{2\rho}\left[\frac{\omega^2\phi^2 + \sigma^2x^2}{\omega^2 + (\sigma(1-I))^2}\right] \tag{D.7}$$

Completing the square for the term

$$\mu^2 - 2\frac{\left(\omega^2\phi + \sigma^2(1-I)x\right)}{\omega^2 + (\sigma(1-I))^2}\mu,$$

we find

$$
\begin{aligned}
&\mu^2 - 2\frac{\left(\omega^2\phi + \sigma^2(1-I)x\right)}{\omega^2 + \left(\sigma(1-I)\right)^2}\mu \\
&= \mu^2 - 2\frac{\left(\omega^2\phi + \sigma^2(1-I)x\right)}{\omega^2 + \left(\sigma(1-I)\right)^2}\mu + \frac{\left(\omega^2\phi + \sigma^2(1-I)x\right)}{\omega^2 + \left(\sigma(1-I)\right)^2} - \frac{\left(\omega^2\phi + \sigma^2(1-I)x\right)}{\omega^2 + \left(\sigma(1-I)\right)^2} \qquad \text{(D.8)} \\
&= \left(\mu - \frac{\left(\omega^2\phi + \sigma^2(1-I)x\right)}{\omega^2 + \left(\sigma(1-I)\right)^2}\right)^2 - \frac{\left(\omega^2\phi + \sigma^2(1-I)x\right)}{\omega^2 + \left(\sigma(1-I)\right)^2}.
\end{aligned}
$$

Substituting the last line of (D.8) for $\mu^2 - 2\frac{\left(\omega^2\phi + \sigma^2(1-I)x\right)}{\omega^2 + \left(\sigma(1-I)\right)^2}\mu$ in expression (D.7), we may now write

$$
\begin{aligned}
A &= \frac{-1}{2\rho}\left(\mu - \frac{\left(\omega^2\phi + \sigma^2(1-I)x\right)}{\omega^2 + \left(\sigma(1-I)\right)^2}\right)^2 - \frac{1}{2\rho}\left[\left[\frac{\left(\omega^2\phi + \sigma^2(1-I)x\right)}{\omega^2 + \left(\sigma(1-I)\right)^2}\right]^2 - \frac{\omega^2\phi^2 + \sigma^2 x^2}{\omega^2 + \left(\sigma(1-I)\right)^2}\right] \\
&= \frac{-1}{2\rho}\left(\mu - \frac{\left(\omega^2\phi + \sigma^2(1-I)x\right)}{\omega^2 + \left(\sigma(1-I)\right)^2}\right)^2 + C(x)
\end{aligned}
$$

Returning to expression (D.6) we may write

$$
\begin{aligned}
f_X(x) &= \int_{-\infty}^{\infty} \frac{1}{\sqrt{2\pi\sigma^2\omega^2}} e^{-\left[\frac{(x-\mu(1-I))^2}{2\sigma^2} + \frac{(\mu-\phi)^2}{2\omega^2}\right]} dx\,d\mu \\
&= \frac{1}{\sqrt{2\pi\left(\omega^2 + \left(\sigma(1-I)\right)^2\right)}} e^{C(x)} \int_{-\infty}^{\infty} \frac{1}{\sqrt{2\pi\frac{\sigma^2\omega^2}{\omega^2 + \left(\sigma(1-I)\right)^2}}} e^{\frac{-1}{2\frac{\sigma^2\omega^2}{\omega^2+(\sigma(1-I))^2}}\left(\mu - \frac{\left(\omega^2\phi + \sigma^2(1-I)x\right)}{\omega^2 + \left(\sigma(1-I)\right)^2}\right)^2} du \\
&= \frac{1}{\sqrt{2\pi\left(\omega^2 + \left(\sigma(1-I)\right)^2\right)}} e^{C(x)}.
\end{aligned}
$$

because the integrand is the probability density function of a normally distributed random variable with mean

$$\frac{\left(\omega^2\phi+\sigma^2(1-I)x\right)}{\omega^2+\left(\sigma(1-I)\right)^2}$$

and variance $\sigma^2\omega^2 \Big/ \omega^2+\left(\sigma(1-I)\right)^2$. It now remains to identify $C(x)$.

$$-2\rho C(x)=\left[\frac{\omega^2\phi^2+\sigma^2x^2}{\omega^2+\left(\sigma(1-I)\right)^2}-\frac{\left(\omega^2\phi+\sigma^2(1-I)x\right)}{\omega^2+\left(\sigma(1-I)\right)^2}\right]^2$$

$$=\frac{\left(\omega^2\phi^2+\sigma^2x^2\right)\left(\omega^2+\left(\sigma(1-I)\right)^2\right)}{\left[\omega^2+\left(\sigma(1-I)\right)^2\right]^2}-\frac{\left(\omega^2\phi+\sigma^2(1-I)x\right)^2}{\left[\omega^2+\left(\sigma(1-I)\right)^2\right]^2}$$

Further simplification reveals

$$-2\rho C(x)=\frac{\left(\omega^4\phi^2+\sigma^2\omega^2x^2+\omega^2\phi^2\left(\sigma(1-I)\right)^2+\sigma^2x^2\left(\sigma(1-I)\right)^2\right)}{\left[\omega^2+\left(\sigma(1-I)\right)^2\right]^2}$$

$$-\frac{\left(\omega^4\phi^2+2\omega^2\phi\sigma^2(1-I)x+\left(\sigma^2(1-I)x\right)^2\right)}{\left[\omega^2+\left(\sigma(1-I)\right)^2\right]^2}.$$

Cancellation of terms produces

$$C(x) = \frac{-1}{2\left(\omega^2 + (\sigma(1-I))^2\right)}\left(x^2 - 2\phi(1-I)x + \phi^2(1-I)^2\right)$$
$$= \frac{-1}{2\left(\omega^2 + (\sigma(1-I))^2\right)}(x - \phi(1-I))^2.$$

Thus, the probability function of X is

$$f_X(x) = \frac{1}{\sqrt{2\pi\left(\omega^2 + (\sigma(1-I))^2\right)}} e^{\frac{-1}{2\left(\omega^2 + (\sigma(1-I))^2\right)}(x-\phi(1-I))^2}.$$

a function that we recognize as a normal distribution with mean $\phi(1-I)$ and variance $\omega^2 + (\sigma(1-I))^2$.

For the setting that was discussed and elaborated in Chapter Eight, $\phi = \sqrt{E/4a^2}\, b$, $\sigma = (1-I)$, $\omega^2 = \sqrt{E/4a^2}$. Thus, for b = 1 a = 4, and E = 658, the mean of X is $3.21(1-I)$, its variance is $3.21 + (1-I)^4$.

D.1 Application to Predictive Boundaries

To compute the predictive power using this prior distribution $\pi(\theta)$ and $\pi(\mu)$ we recall that the investigators choose to compute the predictive power of stopping at information time I_1 based on the value of the test statistic $TS(I_1) = s_1$. They would compute

$$\boldsymbol{P}\left[TS(1) \geq 1.96 | TS(I_1) = s_1\right]$$
$$= \boldsymbol{P}\left[B(1) \geq 1.96 \mid B(I_1) = \sqrt{I_1}s_1\right] = \boldsymbol{P}\left[B(1-I_1) \geq 1.96 - \sqrt{I_1}s_1\right]$$
$$= \boldsymbol{P}\left[N(\mu(1-I_1), (1-I_1)) \geq 1.96 - s_1\sqrt{I_1}\right].$$

The right hand side of the previous expression is a computation that would be a function of the drift parameter μ. However, we know that μ has a prior distribution. The work from the previous section demonstrates that μ is normally distributed with mean $\sqrt{E/4a^2}\, b(1-I)$ and variance $E/4a^2 + (1-I)^4$. Thus, we may complete the conditional power computation by writing

$$
\begin{aligned}
&\boldsymbol{P}\left[TS(1)\geq 1.96 \mid TS(I_1)=s_1\right] \\
&=\boldsymbol{P}\left[N\left(\mu(1-I_1),(1-I_1)\ \right)\geq 1.96-s_1\sqrt{I_1}\right] \\
&=\boldsymbol{P}\left[N\left(\sqrt{\frac{E}{4a^2}}b(1-I),\frac{E}{4a^2}+(1-I)^4\ \right)\geq 1.96-s_1\sqrt{I_1}\right].
\end{aligned}
$$

Continuing, we can write

$$
\begin{aligned}
&=\boldsymbol{P}\left[N(0,1)\geq\frac{1.96\ -\ s_1\sqrt{I_1}\ -\ \sqrt{\frac{E}{4a^2}}b(1-I_1)}{\sqrt{\frac{E}{4a^2}+(1-I_1)^4}}\right] \\
&=1-F_Z\left[\frac{1.96\ -\ s_1\sqrt{I_1}\ -\ \sqrt{\frac{E}{4a^2}}b(1-I_1)}{\sqrt{\frac{E}{4a^2}+(1-I_1)^4}}\right].
\end{aligned}
$$

And the predictive power is a function of the value of test statistic s_1 at information time I_1, and values of the parameters a and b in the prior distribution $\pi(\theta)$. Recall that we find the boundary value of the test statistic d at information time I_1 by solving

$$
CP(H_0)=\gamma=\boldsymbol{P}\left[TS(1)\geq Z_{1-\alpha/2} \mid TS(I)=d\right] \tag{D.9}
$$

As we saw in Chapter Five, we need to solve for the value of d, we first convert expression (D.9) to a statement involving a Brownian motion event,

$$
\gamma=\boldsymbol{P}\left[B(1)\geq Z_{1-\alpha/2} \mid B(I)=\sqrt{I}d\right]
$$

Recognizing this as Brownian motion conditioned on the past we can write

$$
\gamma=\boldsymbol{P}\left[B(1-I)\geq Z_{1-\alpha/2}-\sqrt{I}d\right],
$$

which now can be written as

$$
\gamma=\boldsymbol{P}\left[N\left(u(1-I),(1-I)\right)\ \geq\ Z_{1-\alpha/2}-\sqrt{I}d\right]. \tag{D.10}
$$

Using the Bayesian perspective, we know that μ has a prior distribution associated with it, which in our current paradigm, is a normal distribution. We demonstrated earlier in this appendix that when the values of μ are averaged out, that expression (D.10) continues to follow a normal distribution mean $\sqrt{E/_{4a^2}}\,b(1-I)$ and variance $E/_{4a^2}+(1-I)^4$. Thus, expression (D.10) becomes.

$$\gamma = \boldsymbol{P}\left[N\left(\sqrt{\frac{E}{4a^2}}b(1-I),\left(\frac{E}{4a^2}+(1-I)^2\right)(1-I)\right) \geq Z_{1-\alpha/2}-\sqrt{I}d\right].$$

Simplification yields

$$\gamma = \boldsymbol{P}\left[N\left(\sqrt{\frac{E}{4a^2}}b(1-I),\left(\frac{E}{4a^2}+(1-I)^4\right)\right) \geq Z_{1-\alpha/2}-\sqrt{I}d\right]$$

$$\gamma = \boldsymbol{P}\left[N(0,1) \geq \frac{Z_{1-\alpha/2}-\sqrt{I}d-\sqrt{\frac{E}{4a^2}}b(1-I)}{\sqrt{\left(\frac{E}{4a^2}+(1-I)^4\right)}}\right].$$

Because we know that $\boldsymbol{P}\left[N(0,1)\geq Z_{1-\gamma}\right]=\gamma$, we can write

$$\gamma = \frac{Z_{1-\alpha/2}-\sqrt{I}d-\sqrt{\frac{E}{4a^2}}b(1-I)}{\sqrt{\left(\frac{E}{4a^2}+(1-I)^4\right)}}.$$

It simply remains to solve for the boundary value d,

$$\gamma = \frac{Z_{1-\alpha/2}-\sqrt{I}d-\sqrt{\frac{E}{4a^2}}b(1-I)}{\sqrt{\left(\frac{E}{4a^2}+(1-I)^4\right)}}.$$

$$\gamma\sqrt{\left(\frac{E}{4a^2}+(1-I)^4\right)} = Z_{1-\alpha/2}-\sqrt{I}d-\sqrt{\frac{E}{4a^2}}b(1-I)$$

$$d = \frac{Z_{1-\alpha/2}-\gamma\sqrt{\left(\frac{E}{4a^2}+(1-I)^4\right)}-\sqrt{\frac{E}{4a^2}}b(1-I)}{\sqrt{I}}.$$

And we observe that the boundary value is a function of the conditional power γ, the type I error at the end of the study α, the information time at which the test statistic will be assessed I, and the parameters a and b of the prior distribution for the therapy's effectiveness $\pi(\theta)$.

Appendix E

Standard Normal Probabilities

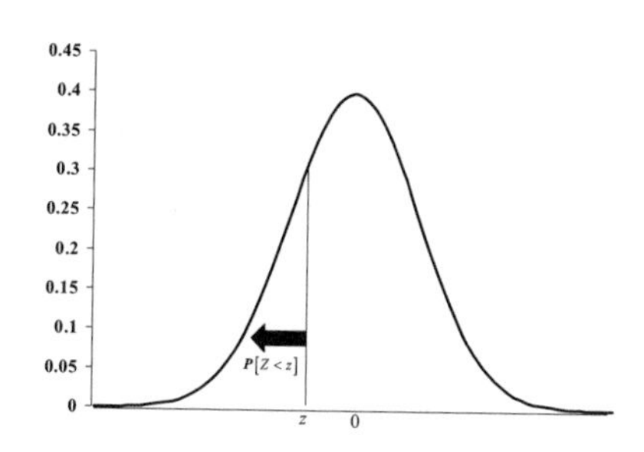

z	P(Z < z)	z	P(Z < z)	z	P(Z < z)	z	P(Z < z)
-3.00	0.001	-1.55	0.061	-0.10	0.460	1.35	0.911
-2.95	0.002	-1.50	0.067	-0.05	0.480	1.40	0.919
-2.90	0.002	-1.45	0.074	0.00	0.500	1.45	0.926
-2.85	0.002	-1.40	0.081	0.05	0.520	1.50	0.933
-2.80	0.003	-1.35	0.089	0.10	0.540	1.55	0.939
-2.75	0.003	-1.30	0.097	0.15	0.560	1.60	0.945
-2.70	0.003	-1.25	0.106	0.20	0.579	1.65	0.951
-2.65	0.004	-1.20	0.115	0.25	0.599	1.70	0.955
-2.60	0.005	-1.15	0.125	0.30	0.618	1.75	0.960
-2.55	0.005	-1.10	0.136	0.35	0.637	1.80	0.964
-2.50	0.006	-1.05	0.147	0.40	0.655	1.85	0.968
-2.45	0.007	-1.00	0.159	0.45	0.674	1.90	0.971
-2.40	0.008	-0.95	0.171	0.50	0.691	1.95	0.974
-2.35	0.009	-0.90	0.184	0.55	0.709	2.00	0.977
-2.30	0.011	-0.85	0.198	0.60	0.726	2.05	0.980
-2.25	0.012	-0.80	0.212	0.65	0.742	2.10	0.982
-2.20	0.014	-0.75	0.227	0.70	0.758	2.15	0.984
-2.15	0.016	-0.70	0.242	0.75	0.773	2.20	0.986
-2.10	0.018	-0.65	0.258	0.80	0.788	2.25	0.988
-2.05	0.020	-0.60	0.274	0.85	0.802	2.30	0.989
-2.00	0.023	-0.55	0.291	0.90	0.816	2.35	0.991
-1.95	0.026	-0.50	0.309	0.95	0.829	2.40	0.992
-1.90	0.029	-0.45	0.326	1.00	0.841	2.45	0.993
-1.85	0.032	-0.40	0.345	1.05	0.853	2.50	0.994
-1.80	0.036	-0.35	0.363	1.10	0.864	2.55	0.995
-1.75	0.040	-0.30	0.382	1.15	0.875	2.60	0.995
-1.70	0.045	-0.25	0.401	1.20	0.885	2.65	0.996
-1.65	0.049	-0.20	0.421	1.25	0.894	2.70	0.997
-1.60	0.055	-0.15	0.440	1.30	0.903	2.75	0.997

Index

C

D

E

F

G

R

S

T

U

W

X